Oxford Illustrated Computing Dictionary

Compiled by
Alison Page
Andrew Delahunty

Great Clarendon Street, Oxford, OX2 6DP, United Kingdom

Oxford University Press is a department of the University of Oxford. It furthers the University's objective of excellence in research, scholarship, and education by publishing worldwide. Oxford is a registered trade mark of Oxford University Press in the UK and in certain other countries

Images courtesy of Shutterstock
Cover illustration by Jenny Wren

British Library Cataloguing in Publication Data

Data available

ISBN: 978 0 19 277245-9
10 9 8 7 6 5 4 3 2 1

Printed in China

Paper used in the production of this book is a natural, recyclable product made from wood grown in sustainable forests. The manufacturing process conforms to the environmental regulations of the country of origin.

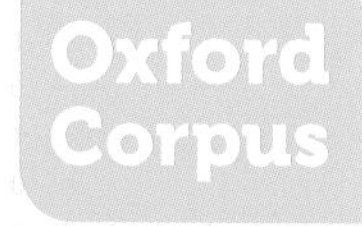

You can trust this dictionary to be up to date, relevant and engaging because it is powered by the Oxford Corpus, a unique living database of children's and adults' language.

Contents

Introduction

The *Oxford Illustrated Computing Dictionary* is a comprehensive quick reference guide to computing vocabulary for both the classroom and the home. It contains over 1000 computing words and terms from the curriculum, national and international, along with words in everyday use and technical vocabulary.

catch words
show the first and last word on the page and guide you to the correct place to find the word you need

headword
is in alphabetical order, in blue

alphabet
the alphabet is given on every page with the letter you are in highlighted so you can find your way around the dictionary easily

alternative words
which can be used instead of the headword to mean the same thing

definition
shows what the word means and if a word has more than one meaning, then each meaning is numbered

derivative
shows you an additional word from the same family as the headword

A B C D E F G H I J K L M N O P Q R S T U V W X Y Z

CAD
CAD stands for *Computer Aided Design*. It means making the design of an object on the computer screen. The design can then be passed on to CAM.

see also CAD/CAM, CAM

CAD/CAM
CAD/CAM is combining CAD and CAM into a complete process.

see also CAD, CAM

CAL (*also* CBT, Computer Based Training)
CAL stands for *Computer Assisted Learning*, the use of computers to help teach students. The computers can run tests and mark the answers.

calculate
To calculate is to work out the value of a numerical expression.

> calculation
You do a calculation when you work out the answer to a numerical expression. You can set out a calculation using a formula.

see also expression, formula

call
To call a subroutine (either a function or a procedure) means to put the name of the subroutine in a program. Then all the commands in the subroutine will be carried out.

see also subroutine, function, procedure

CAM
CAD stands for *Computer Aided Manufacture*. The computer takes the design of an object and then controls a device that makes the object, such as a 3D printer or a cutter.

see also CAD, CAD/CAM

cancel button
A cancel button may be part of a dialogue box. You click on the cancel button to close the dialogue and stop the action. You will go back to where you were before.

see also dialogue box

canvas
A canvas is the background to a graphic image. Graphic objects can be arranged on the canvas.

see also object

Caps Lock
The Caps Lock key is on the keyboard. When Caps Lock is on all your typing will be in capitals (upper case). When Caps Lock is off your typing will be in lower case. Sometimes there is a light on the keyboard to remind you that Caps Lock is on.

see also keyboard, toggle, case

carry bit
When you do sums you sometimes have to 'carry one'. You have to do the same thing with binary numbers. The carry value is called a carry bit.

case
Case means whether text is in capitals or not. Upper case means text is in CAPITAL LETTERS. Other text is called lower case.

18

Definitions explain terms in full and are easy to understand. Where a word has several meanings, different meanings are numbered and often other related words are listed. This is a great way to build vocabulary. Green and yellow panels bring together in one place words that are related to the headword or can be used with it to talk or write about a topic. Illustrations also help to explain the meaning.

The thematic supplement explores in more detail some of the key computing terms and concepts in focus areas, including programming and computer games.

cashpoint to CGI

WORD BUILD

> case-sensitive

Sometimes an input (such as a password) is case-sensitive. This means that it matters whether you input upper or lower case.

cashpoint (*also* ATM)
A cashpoint is a bank outlet. It is controlled by a processor. After you enter a bank card and type a key code, you can get money from your account.

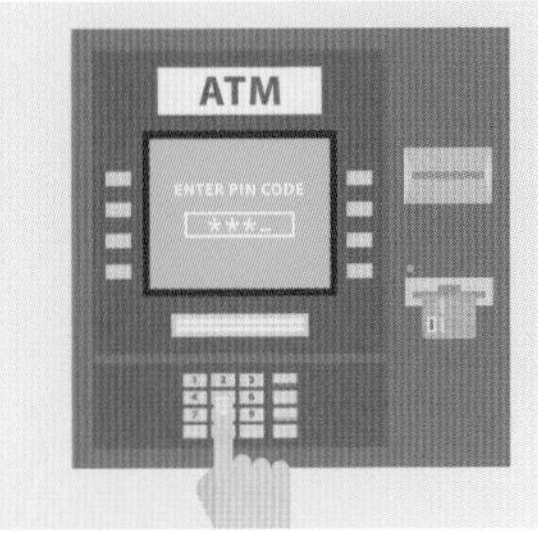

catfishing
Catfishing is a name for a type of danger on social media. Catfishing is when someone pretends to be someone else, for example when an adult pretends to be a child.

see also social media

⚠ WATCH OUT!

Sometimes catfishing is used to trick people into meeting up in real life. Do not agree to meet up with people you do not know.

OK

CBT *see* CAL

CD *see* optical storage

CD drive
A CD drive is a piece of hardware that lets the computer read the data from a CD.

see also optical storage

CD-RW *see* optical storage

cell
A cell is the smallest part of a spreadsheet. It is where a row crosses a column. A cell can hold one piece of data.

see also spreadsheet, row, column

cell phone *see* mobile phone

cell reference
A cell reference identifies a cell in a spreadsheet. It is the column letter followed by the row number, for example B6. You can put a cell reference into a formula. The formula will use the data from that cell.

see also spreadsheet, formula

○ absolute cell reference

An absolute cell reference is fixed. It will not change if you copy the formula to a new cell.

○ relative cell reference

A relative cell reference is not fixed. When you copy the formula, the reference will change. This lets you copy a formula down a whole column or across a whole row of a table. Each version of the formula will use data from a different cell.

CGI
CGI stands for *Computer Generated Imagery*. Special effects can be added to a film by CGI.

a b c d e f g h i j k l m n o p q r s t u v w x y z

19

Word Build panel
shows other words which work together with the headword in the same topic area

other related words
point you to other words that help to explain this word or build more knowledge in this topic area

related words panel
shows words which are linked to the headword in form and meaning

warning note
gives additional information about what you need to watch out for

illustration
helps to show the meaning of the word

cross-reference
points you to the main entry which gives you the definition of the word

Symbols

@ (*also* at)

The @ symbol is used in social media. It is pronounced 'at'. The @ symbol marks a username and is also used in an email address.

see also social media

.com (*also* dot com)

.com is pronounced 'dot com'. The address of a business website often ends with the letters .com. A 'dot com' company is one that makes money through the internet.

see also TLD, e-commerce

(*also* hashtag)

The # symbol is used in social media. It is pronounced 'hashtag'. You can put # in front of a word or phrase, for example #FunnyPets. The hashtag tells people the topic of your post. Someone who is interested in a topic can look for it by using the hashtag. They will find posts on that topic.

see also social media, post

Numbers

2-in-1 device

A 2-in-1 device is a tablet that converts to a laptop. You can stand it upright in a holder and you can attach a keyboard.

see also laptop, tablet

32-bit

Computer data is stored in bits. 32-bit is a measure of processor word size (how powerful the processor is). Some processors have a 32-bit word size.

see also word size

3D

3D stands for *three dimensions* or *three-dimensional.* A 3D experience is not flat. You can experience depth or distance.

3D barcode *see* barcode

3D printer

A 3D printer builds up a solid object from a substance such as plastic, resin or metal. The substance is added in blobs that set or in grains that are glued together.

see also CAD/CAM

3G, 4G, 5G

3G, 4G and 5G are types of mobile phone connection. The letter G stands for *Generation.* The higher the number, the more modern and quicker the connection is.

404 message

A 404 message is a type of error message. It is pronounced 'four-oh-four'. You may see this message in the web browser window. It tells you that the browser could not load a web page, for example because the web page is no longer available.

see also browser, load, web page

64-bit

Computer data is stored in bits. 64-bit is a measure of processor word size (how powerful the processor is). Some processors have a 64-bit word size.

see also word size

abort

To abort a process means to stop it. The process will halt and any actions will be reversed.

absolute cell reference *see* cell reference

abstraction

Abstraction means making a simple version of a more complicated system. The simpler version is called an abstract model. To make the simple model you leave out some of the details. You only keep the details that are related to the task.

see also model

Access

Access is the name of database software made by Microsoft.

see also database, Microsoft

access

If a computer can access something, it can get to it because there is a working connection. For example, if a computer can access a storage area, it can read or write data to that storage area.

see also read, write, remote access

account

An account is a way of giving people their own personal access to a computer system. This is important in a shared system used by many people, for example a network or a social media site. Usually an account needs a username and password.

see also username, password

user account

A user account is the account of an ordinary user. When you 'log in' to a system you connect to your user account. Then you can access your files or entries.

see also log in, log on

admin account

An admin account is a type of user account. If you log in with an admin account you have admin privilege. That means you have more powers than an ordinary user.

see also admin privilege

accumulator *see* ALU

Acrobat

Acrobat is the name of a group of software applications made by a company called Adobe. The applications let you make and view documents. The application for viewing documents is free. The documents are made in a format called PDF.

see also PDF

action

An action by a computer is usually called an operation.

see also operation

activate

To activate something means to make it start working. Some software cannot be used until it is first activated. You may need a product key (a kind of password) which the company sends to your email address. This is to prevent people using illegal copies of the software.

active window *see* window

A B C D E F G H I J K L M N O P Q R S T U V W X Y Z

add-on

An add-on is an extra piece of software added to an app. Add-ons let the app do extra tasks.

see also app

address *see* email, IP address, memory address, web address

address bar

An address bar appears at the top of a web browser. You can type a web address (URL) in the address bar. The web browser looks for that page which then appears in the main window of the browser.

see also browser, web page, URL

admin (*also* administrator)

Admin means the person who is in charge of a computer system. This person often has a special login password.

see also login, admin privilege, system administrator

admin account *see* account

admin privilege

Admin privilege means the extra powers that the admin has. These let the admin make changes to a computer system.

see also admin, system administrator

Adobe Flash *see* Flash

ADSL

ADSL stands for *asymmetric digital subscriber line.* ADSL is a way to use an ordinary phone line to connect to the internet. The connection is quite slow.

see also download, upload

AI

AI stands for *artificial intelligence*. This means making a computer that can solve problems like a person can. AI is an aim of computing but nobody has fully achieved it yet.

algorithm

An algorithm is a set of rules that shows the steps needed to solve a problem. An algorithm can be used as the plan for a computer program.

see also flowchart, pseudocode

aligned (*also* justified)

If the text of a paragraph is aligned, it lines up neatly so that the margin is neat and straight. We can also say the text is justified. The most common type of alignment is left-aligned (also called left-justified). Other types of alignment include right-aligned (or right-justified), centred in the line, and aligned to the top or bottom of the document.

> alignment

Alignment is the way that the text of a paragraph is aligned.

WORD BUILD

> fully justified

If text is fully justified, both the right and left margins are straight. The computer makes this happen by varying the space in between the words in the line.

Alt key (*also* option key)

The Alt key is one of the keys on a normal keyboard. It is marked 'Alt', short for *Alternate*. You press the Alt key together with another key (as you do with the Shift key). This makes a keyboard shortcut command. The equivalent on a Macintosh computer is the 'option' key.

keyboard shortcut

ALU

ALU stands for *arithmetic and logic unit*. This is the area of the processor where calculations are carried out.

see also process, CPU

WORD BUILD

> accumulator

In a processor, the accumulator is a small area of electronic memory inside the ALU. It stores the result of the current task.

Amazon

Amazon is a widely-used e-commerce site. Amazon customers can buy things from the Amazon website.

see also e-commerce

analogue *also* analog

Data can be analogue data or digital data. Analogue data does not exist as exact number values, but varies smoothly between different levels. For example, temperature is an analogue measurement. It does not jump from 10°C to 11°C, but instead rises smoothly through all the temperatures in between. You cannot show all the exact values, even if you use a decimal point.

see also digital

analogue-to-digital converter

Analogue data and digital data are different. The computer can only process digital data, so analogue data must be turned into digital data. This is done by using an analogue-to-digital converter.

AND gate

In a processor, an AND gate is a type of electronic gate. Two electrical signals go into the gate. One signal comes out. If both inputs are ON the output is ON.

see also gate, process

AND operator *see* logical operator

Android

Android is the name of an operating system made by Google. It is mainly used on mobile devices like phones and tablets.

see also operating system

android

An android is a robot that looks and acts just like a person. Fully realistic androids have not yet been invented.

angle brackets *see* brackets

A B C D E F G H I J K L M N O P Q R S T U V W X Y Z

antispam

Antispam is software that stops spam email getting into your inbox.

see also spam, inbox

antivirus (*also* anti-virus)

Antivirus is software that finds and removes viruses from your computer. There are many companies that make antivirus software. Some of this software is free.

see also virus, malware

app (*also* application software)

App is short for *application* or *application software*. An app usually does one thing. You can download an app to any computer, phone or tablet. Then you can use the app on that device. Some apps only work with specific operating systems.

see also software

Apple

Apple is the name of a US technology company that makes a wide range of electronic devices such as Macs, iPhones and iPads.

applet

An applet is a small software application. An applet is often part of a web page and does a particular job, such as playing an embedded video. Applets are often written in the programming language Java.

see also app, embedded video

application software *see* software

architecture

Computer architecture is how the parts of a computer are organised. Most computers have a standard architecture.

archive

An archive is a storage area used to keep old versions of files.

argument *see* parameter

arithmetic operator

An arithmetic operator is a symbol that tells the computer to carry out a calculation. Some program commands include arithmetic operators. The main arithmetic operators are: **+ (plus), - (minus), * (multiply), / (divide)**. When we use it in this way, arithmetic is pronounced 'arith-MET-ic'.

see also operator, div, MOD

array *see* data structure

arrow key

The arrow keys are four keys on the keyboard, each marked with an arrow pointing in a different direction. Once an object has been selected, the arrow keys are used to scroll through it or move it about on the screen.

see also keyboard

ASCII

ASCII stands for *American Standard Code for Information Interchange*. ASCII is a number code. All keyboard characters are converted to a number code so computers can work with them. There are 128 codes in the ASCII system. ASCII is part of a much larger code system with more than 65 million codes, called Unicode.

see also Unicode, character

assembly language

Assembly language is a programming language in which most of the commands are only three letters long. The commands match up exactly to similar machine code commands. Assembly language is translated into machine code using a translator called an assembler.

see also code, low-level language, programming language

assign

To assign a value to a variable means to store the value in the variable. In many programming languages the assignment command takes this form: variable = value.

see also variable, value

> assignment

Assignment is the command that makes the computer assign a value to a variable.

WORD BUILD

> assignment operator

The assignment operator is the symbol that makes the computer assign a value to a variable. In many programming languages the assignment operator is the equals sign.

see also operator, equals sign

ATM *see* cash point

attach

If you attach a file to an email, the file will arrive joined to the email.

attachment

An attachment is a file that is attached to something else, such as an email. When you get the email you get the file too.

attribute

An attribute is a data value. An attribute always applies to an object. You cannot have an attribute without an object and the name of the attribute must reflect the property of the object that the value represents. For example, the attribute 'height' is a data value and must reflect the height of an object, a person or a shape.

see also object, variable, assign, data

audio

Audio means to do with sound. An audio file holds sounds in digital form.

augmented reality

Augmented reality is a way of adding extra information to a real-life scene. This is done, for example, by wearing special glasses or on the video display of your phone. The extra information might be labels or symbols, such as street names or directions.

see also VR

authenticate

To authenticate something is to check that it is really what it says it is. For example, a computer may be able to authenticate a user or a website.

see also log in

auto-

The term auto- can be put before actions. It means the action can happen on its own, without a person. For example, an auto-pilot is a system that can fly a plane.

autofill

Autofill is a feature that automatically fills in some data, for example when you are using a spreadsheet. You might start to fill a column with numbers. The computer can complete the column by filling in all the numbers, in the right order.

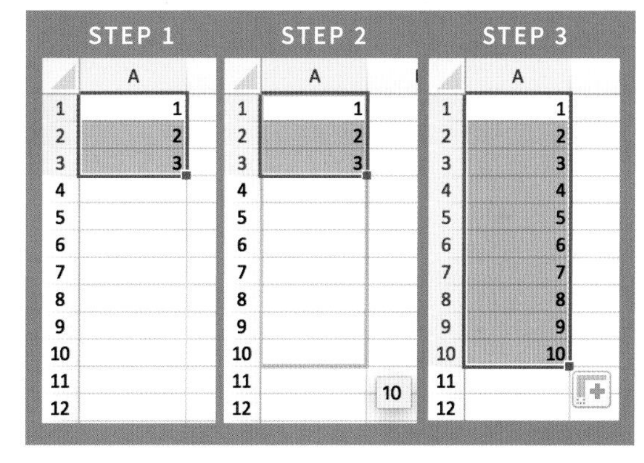

A
B
C
D
E
F
G
H
I
J
K
L
M
N
O
P
Q
R
S
T
U
V
W
X
Y
Z

automate

If a system is automated, it is controlled entirely by a computer. A human operator is not needed. Control systems are used to automate processes.

see also control device

› automation

Automation means a process is controlled entirely by a computer. A human operator is not needed.

autoplay

Autoplay means something will play on its own. An autoplay video is a video in a website which starts to play when you open the website.

see also embedded video

AutoSum

AutoSum is a spreadsheet feature that automatically adds up all the numbers in a row or column.

see also sum, spreadsheet

avatar

An avatar is your user identity in social media or a game. You might choose a username that is not your name or choose a picture that is not a picture of you.

see also social media

axis (*plural* axes)

An axis is a line along the side of a graph or an image. It shows number values or coordinates. The x-axis marks the horizontal edge and the y-axis marks the vertical edge.

see also coordinates

backed

If printing is backed, the printing goes on both sides of the page.

see also printer

background

The desktop wallpaper is sometimes called the background.

see also desktop

› in the background

Apps and applications can run 'in the background'. That means they do not show on the screen.

backspace key

The backspace key is on the keyboard. It is often at the top right and has a backwards arrow on it. It deletes the character before the cursor.

see also key, keyboard, delete key, cursor

backup (*also* back-up)

A backup is an extra copy of a file that you make and store somewhere safe. If your main copy is lost you can restore the file from the backup.

see also restore, corrupt

› back up

If you back up a file, you make an extra copy of it and store it somewhere safe.

bandwidth (*also* connection speed)

Bandwidth means the maximum possible speed of a communication link. A link with a high bandwidth can send data more quickly than one with a low bandwidth.

barcode

A barcode is a way of recording a code number. The number is shown using printed black and white stripes. The computer can read the barcode with a scanner. Barcodes are used in shops to record product codes.

3D barcode

A 3D barcode is a larger kind of barcode that can store more information. It often stores the address of a website. You can get an app on your phone that will read a 3D barcode like this and then open the website.

barcode reader

A barcode reader is an input device that scans a barcode. It sends the code number to the computer.

see also barcode

base two, ten, sixteen

see number base

baud rate

Baud rate is a measure of the speed of a comms link. It tells you the number of symbols sent in one second. Digital signals use only two symbols, the bits 1 and 0. So for computers, baud rate is the same as bandwidth and is measured in bits per second (bps).

see also bit, bps, bandwidth

bespoke *see* customised

beta test

All software is tested before it is offered for sale to users. A beta test is when the users help with this testing, by trying out the software and saying if they find any problems.

see also testing

binary

Binary means made of two different things. Binary numbers are made of 1s and 0s. Binary numbers are in base 2.

see also digital, hexadecimal, number base

binary code

Binary code is a system of representing all types of data and instructions in a computer by means of binary numbers (1s and 0s).

binary search *see* search algorithm

binary shift

Binary shift is a way to multiply binary numbers.

binary signals

Binary signals are used for computer communications. The signals represent the numbers 1 and 0.

Bing

Bing is the name of a search engine provided by Microsoft. It is free to use.

see also search engine

biometrics

Biometrics is a way of checking user identity by using aspects of the human body for example your face, eyes or fingerprints.

see also authenticate, identity theft

WORD BUILD

> facial recognition

Facial recognition is a way for a computer to check who you are. It uses features that are hard to fake, such as the spacing of eyes.

> retinal scan (*also* iris scan, eye scan)

Your eye does not look exactly like anyone else's. A retinal scan checks who you are by shining a light in your eye to see the patterns. A similar method is called an iris scan. The retina and iris are parts of your eye.

> thumb print (*also* palm print, finger print)

Your thumb print is not like anyone else's. A computer can scan your thumb to check who you are.

⚠ WATCH OUT!

Do not worry - people cannot cut off a bit of you to fool biometrics. That wouldn't work.

OK

BIOS

BIOS stands for *Basic Input-Output system*. The BIOS is the software that controls the computer start-up. BIOS is stored in ROM.

see also ROM, boot, start up

bit

Bit is short for *binary digit*. A bit is a single digit 1 or 0. All binary data is made of bits.

see also byte, binary

bitcoin

Bitcoin is an online money system that is not controlled by any government. It is based on prime numbers and uses blockchain technology. The value of bitcoins can change a lot.

see also blockchain

bitmap graphics

Bitmap graphics is a way of storing an image in digital form. An image is made of tiny points called pixels. The colour of each pixel is stored using a colour code.

see also pixel, vector graphics

WORD BUILD

> bitmap image

A bitmap image is an image made using bitmap graphics.

bits per second *see* bps

block-based language *see* visual programming language

blockchain

Blockchain technology is a way of recording a series of transactions. It is encrypted (put into a secret code). Bitcoin transactions are stored using blockchain.

see also bitcoin, transaction, encrypt

blog (*also* web log)

Blog is short for *web log*. It is a website that you make yourself and on which you write about your own life or opinions. Other people can read the blog and many blogs allow comments from readers.

see also social media

> blogger

A blogger is a person who writes a blog.

Bluetooth

Bluetooth is a way to link devices so that they can share data. The devices must be close together. The link is wireless. For example, you can connect Bluetooth headphones to a music player in your pocket.

Blu-ray *see* optical storage

bold

Bold text is text that is made with a thick line. It stands out from other text.

bookmark (*also* favourite)

A bookmark is a way to mark a web page you like. Your web browser has a button for doing this. The web browser saves your bookmarks. The list of bookmarks is a list of your favourite sites, so you can quickly find them again.

see also browser

Boolean

Boolean means to do with logic. A Boolean expression can have the value True or False. The name comes from George Boole, who invented modern logic.

see also expression, data type

boot

To boot a computer means to start it. This is also called 'booting up'. Electricity goes into the computer. The computer loads the operating system and all the devices get ready for you to use them.

see also reboot, operating system, start up

bot (*also* internet bot, web robot)

A bot is a small program that runs on the internet. It often records facts about different web pages. The most common type of bot is a web crawler.

see also web crawler

botnet

A botnet is a group of computers that are connected together and are used to run bots. A botnet can be used for bad purposes, such as sending spam email.

see also bot, spam

bps

bps stands for *bits per second*. This measures the speed of a communications link. A connection with a high bps can send data more quickly than one with a low bps. Bits per second can be measured in Kbps (thousand bits per second) or Mbps (million bits per second). Bps (with a capital B) stands for *bytes per second*.

see also bit, byte, bandwidth

a b c d e f g h i j k l m n o p q r s t u v w x y

brackets

Brackets are symbols used in programming for a variety of purposes. They should always be used in pairs: an open and a close bracket. The expression between the brackets is enclosed by the brackets. Brackets are also used in expressions such as spreadsheet formulas. The bracket symbols can be found on the keyboard.

see also keyboard, expression

angle brackets (*also* braces)

Angle brackets look like this: < >. They are used in programming. For example, in web programming, angle brackets enclose HTML commands.

curly brackets

Curly brackets look like this: { }. They are used in programming. For example, in the C programming language, curly brackets enclose program structures such as loops or functions.

round brackets (*also* parentheses)

Round brackets look like this: ( ). These are also called parentheses (pronounced par-EN-thu-SEEZ). In most programming languages, parentheses are used to enclose expressions such as calculations.

square brackets

Square brackets look like this: [ ]. In many programming languages, square brackets are used to enclose lists and other data structures.

branch

In programming, a branch is part of a branching structure. A branch is one of the two or more choices that follow a logical test. The program will go down one branch or the other.

branching structure

A branching structure is a feature of an algorithm or program. A branching structure starts with a logical test. The computer will carry out different commands, depending on the result of the test. It can also be called a conditional structure.

broadband

Broadband is a strong internet connection that allows information to be sent very quickly. A broadband connection can be wired or wireless. The connection is always turned on whether or not it is being used.

see also bps, bandwidth, connection speed

broadcast

A broadcast signal is used for mass communication and is not directed to one receiver. A radio programme is broadcast, which means that anyone can listen to it. A phone call is not broadcast, which means it is directed to one person only.

broken link

A broken link is a web link that does not work. It is a link to a site that does not exist, perhaps because the site has been shut down.

see also web link, 404 message

browse

To browse means to look at websites using a web browser. You move from page to page using hyperlinks.

see also browser, hyperlink

browser (*also* web browser)

A browser is a software application that allows you to look at web pages. A web page is created using HTML. The HTML comes to your computer through an internet link. The web browser uses the HTML or some other language to create the web page on your computer. It may also need to download files such as images and sounds.

see also web page

browser cookie *see* cookie

buffer

A buffer is a small area of digital memory used to store a signal. A buffer is used to join a fast system to a slower system, for example when watching internet TV. Sometimes it takes a few moments to download a complete file or signal. While the download is underway, the signal is stored in a buffer. You might see the message 'buffering' on the screen. Then, as you start to watch the programme, the rest is copied through to the buffer, so there is no more delay.

bug

A bug is a fault or error in a computer program.

see also debug, error

bullying online *see* cyber-bullying

bus

A bus is an electronic connection. The different parts of the processor are connected by buses. A bus can use serial or parallel transmission.

see also process, parallel, serial

button

A button is part of a dialogue box. It is often an oblong shape with round corners. It has words on it that tell you what the button does. You click on the button (for example, the OK button) to make that happen.

see also dialogue box, OK button

byte

A byte is a group of 8 bits. The computer's memory is organised into bytes.

see also bit, binary, word size

C#

C# is a programming language developed by Microsoft. The name is pronounced 'C sharp'. It is a general purpose language that can be used for all types of programs. It was developed from an earlier language called C.

see also high-level language, programming language, C++

C++

C++ is a programming language. The name is pronounced 'C plus plus'. C++ is suited to object-oriented programming (OOP). It is used for making large complicated software systems.
It was developed from an earlier language called C.

see also high-level language, programming language, C#, OOP

cable

Cable is wire that you can use to send a signal.

cache

A cache is a small area of digital memory. It is used to store data that you might need again. For example, if you visit a particular website a lot, your computer might store data from the site in a web cache on your computer instead of taking data from the site each time.

CAD

CAD stands for *Computer Aided Design*. It means making the design of an object on the computer screen. The design can then be passed on to CAM.

see also CAD/CAM, CAM

CAD/CAM

CAD/CAM is combining CAD and CAM into a complete process.

see also CAD, CAM

CAL (*also* CBT, Computer Based Training)

CAL stands for *Computer Assisted Learning*, the use of computers to help teach students. The computers can run tests and mark the answers.

calculate

To calculate is to work out the value of a numerical expression.

> calculation

You do a calculation when you work out the answer to a numerical expression. You can set out a calculation using a formula.

see also expression, formula

call

To call a subroutine (either a function or a procedure) means to put the name of the subroutine in a program. Then all the commands in the subroutine will be carried out.

see also subroutine, function, procedure

CAM

CAD stands for *Computer Aided Manufacture*. The computer takes the design of an object and then controls a device that makes the object, such as a 3D printer or a cutter.

see also CAD, CAD/CAM

cancel button

A cancel button may be part of a dialogue box. You click on the cancel button to close the dialogue and stop the action. You will go back to where you were before.

see also dialogue box

canvas

A canvas is the background to a graphic image. Graphic objects can be arranged on the canvas.

see also object

Caps Lock

The Caps Lock key is on the keyboard. When Caps Lock is on all your typing will be in capitals (upper case). When Caps Lock is off your typing will be in lower case. Sometimes there is a light on the keyboard to remind you that Caps Lock is on.

see also keyboard, toggle, case

carry bit

When you do sums you sometimes have to 'carry one'. You have to do the same thing with binary numbers. The carry value is called a carry bit.

case

Case means whether text is in capitals or not. Upper case means text is in CAPITAL LETTERS. Other text is called lower case.

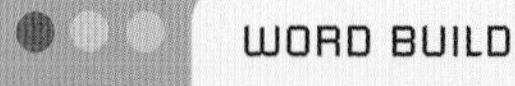

WORD BUILD

> case-sensitive

Sometimes an input (such as a password) is case-sensitive. This means that it matters whether you input upper or lower case.

cashpoint (*also* ATM)

A cashpoint is a bank outlet. It is controlled by a processor. After you enter a bank card and type a key code, you can get money from your account.

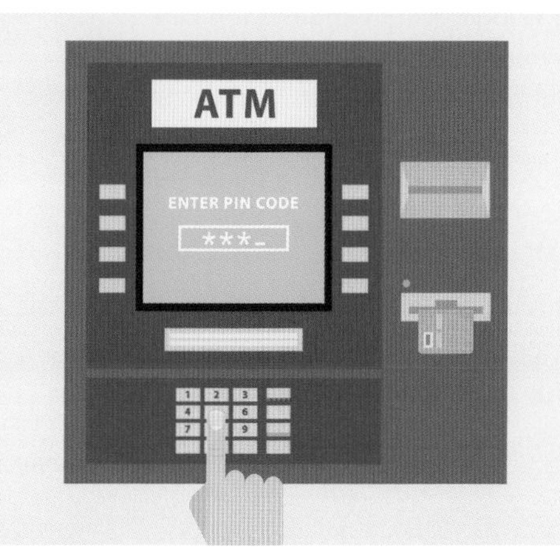

catfishing

Catfishing is a name for a type of danger on social media. Catfishing is when someone pretends to be someone else, for example when an adult pretends to be a child.

see *also* social media

⚠ WATCH OUT!

Sometimes catfishing is used to trick people into meeting up in real life. Do not agree to meet up with people you do not know.

OK

CBT *see* CAL

CD *see* optical storage

CD drive

A CD drive is a piece of hardware that lets the computer read the data from a CD.

see *also* optical storage

CD-RW *see* optical storage

cell

A cell is the smallest part of a spreadsheet. It is where a row crosses a column. A cell can hold one piece of data.

see *also* spreadsheet, row, column

cell phone *see* mobile phone

cell reference

A cell reference identifies a cell in a spreadsheet. It is the column letter followed by the row number, for example B6. You can put a cell reference into a formula. The formula will use the data from that cell.

see *also* spreadsheet, formula

absolute cell reference

An absolute cell reference is fixed. It will not change if you copy the formula to a new cell.

relative cell reference

A relative cell reference is not fixed. When you copy the formula, the reference will change. This lets you copy a formula down a whole column or across a whole row of a table. Each version of the formula will use data from a different cell.

CGI

CGI stands for *Computer Generated Imagery*. Special effects can be added to a film by CGI.

A B C D E F G H I J K L M N O P Q R S T U V W X Y Z

character

1 A character is a single text symbol. Letters, digits and punctuation marks are examples of text characters. Every character has a Unicode value.

see also Unicode, ASCII

2 In gaming, a character is an individual in a computer game. Characters can complete actions such as jumping or fighting. They can be people, animals or monsters. Not all computer games have characters. For example, a puzzle game like Tetris does not have characters.

see also NPC

chat (*also* online chat)

Online chat means a conversation over the internet. It can be with text or live video. It can be to one person or many. You see the message right away and you can respond right away.

see also social media

checkbox (*also* check box, tick box)

A checkbox is one of the small boxes found on an online form. You click a checkbox to choose an option. The boxes you click will be marked with a tick (or other mark).

see also online form, dialogue box

chip (*also* microchip, silicon chip)

A chip is an electronic component, usually made of silicon. Microscopic electronic circuits are etched (carved with acid) on the surface of the chip.

see also microprocessor, circuit board

Chip and PIN (*also* chip and pin)

Chip and PIN is a way to store identity on a card, for example a bank card. The chip is made of flash memory. PIN stands for *Personal Identity Number*. The computer reads the information on the chip and you enter the PIN using a keypad.

see also contactless payment, keypad

Chrome

Chrome is the name of a web browser made by Google, which is the same company that makes the Google search engine. You can download it for free onto your computer.

see also browser, Google

circuit

A circuit is the complete path that an electric current flows along. Computer processors are made of microscopic circuits.

see also chip, process

circuit board (*also* printed circuit)

A circuit board is a piece of board or card that electrical components are fixed to. Metal channels on the board act as connectors between components.

see also motherboard

click

To click on something means when the mouse pointer is on an object or in the right place on the screen, click the button on the mouse to select it. Many types of mouse have more than one button but click usually means click the left-hand button.

see also mouse

control-click

Control-click means press the control key and click the mouse button at the same time.

double-click

Double-click means click the mouse button twice, quickly.

right-click

Right-click means click the right-hand button on the mouse.

clickable

If something is clickable, you can click on it. A feature on the screen might be clickable, so that when you click on it something will happen.

see also mouse

client *see* client-server

client-server

A client-server system is a way that computers share data. The server computer stores important data. The client computer gets the data from the server.

see also P2P

WORD BUILD

client

A client computer gets data from a server. When your computer uses data taken from a website, your computer is working as a client computer.

client-side processing

The client computer might process the data it gets from the server. This is called client-side processing.

server

A server computer stores data. Other computers use that data. A server is often a large, powerful computer with a lot of storage.

server-side processing

The server computer might process the data. It sends the results back to the client computer. This is called server-side processing. For example, when you buy something from a website, the website computer processes your payment. Then it sends you a message to say the purchase is complete.

clip art

Clip art means small images made with computer graphics. You can paste clip art images into your documents. Many apps come with free clip art images that you can use.

A B C D E F G H I J K L M N O P Q R S T U V W X Y Z

clipboard

The clipboard is a small area of computer memory used in cut and paste. You usually cannot see what is in the clipboard, so you need to remember what you cut or copied.

see also cut and paste

clock speed *(also* processing speed)

Every computer has a clock inside it which controls the timing of the computer. The clock speed controls how fast the computer works.

see also overclocking

close

To close a file means you stop working with the file. It is removed from the computer's electronic memory. If you have made changes you must save the file before closing it. If you do not save, the changes will be lost.

close button

The close button is at the top right of a window. It has a red X on it. You click this button to close the window. This will usually shut the app that is running in the window.

see also window

cloud-based

If something such as a service is cloud-based, you use it over the internet. A common example is cloud-based storage. This allows you to store your files on a remote computer through an internet connection.

see also storage, remote content

CMYK *see* colour

code

In programming, code is the name for the instructions that make up a computer program. There are two types of code: machine code and source code.

machine code

Machine code is a number code. Every instruction has a code number. The computer understands machine code instructions. A file of machine code instructions is called an executable file. If you run the file the computer will carry out the instructions.

see also compiler, executable file, instruction, run, assembly language

source code

Source code is written in a programming language. The computer cannot understand source code. The computer can only understand machine code. Source code has to be turned into machine code so that the computer can understand the instructions. Turning source code into machine code is called translation.

see also compiler, programming language, translate

code library

A code library is a collection of subroutines stored in files. Programmers are often generous and share the subroutines they made so that other programmers can use them in their own programs.

see also subroutine, module

code quality

Code quality means how good source code is. Quality code is free of errors and does the job it is supposed to do. It should be readable and robust.

WORD BUILD

readable

Readable code can be easily read and understood by programmers. That makes it easy to work with. Code is more readable if it has well-chosen identifiers. Comments make code more readable.

see also comment, identifier »

> robust

Robust code does not break down easily, even if users make input errors. Adding validation to a program makes it more robust.

see also validation, testing

collision detection

Collision detection helps a robotic system to move about. The robot can tell if it touches anything and can move out of the way.

see also Roomba, robot

colour

A computer stores colours using code numbers. There is more than one way to code colours.

WORD BUILD

> CMYK

CMYK is a colour code system. CMYK stands for *Cyan, Magenta, Yellow, Key* ('key' means black). These are the four colours of ink in colour printing. The amount of each colour is stored as a number.

> colour depth

Colour depth tells you how many colour codes are available. A larger number of codes results in greater colour depth, which means that a bigger range of colours can be used to make an image.

> colour palette

A colour palette is the set of colours used in an image.

> RGB

RGB is a colour code system. RGB stands for *Red, Green, Blue*. These are the three colours of light used to make a coloured dot on screen. The amount of each colour is stored as a number.

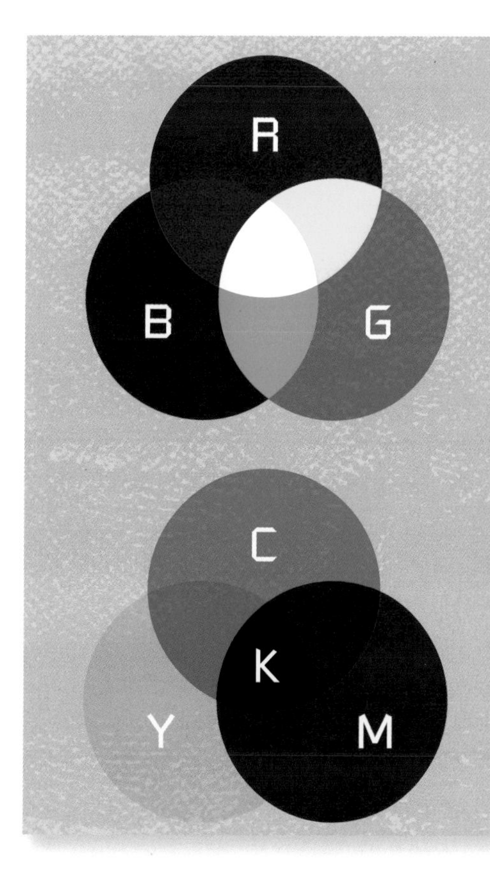

colour depth *see* colour

colour palette *see* colour

column

A column is a vertical group of cells. Spreadsheets and data tables are organised into columns and rows.

see also data table, spreadsheet, row

.com *see page 6*

command

A command is a part of a computer program. It tells the computer to do one action (operation). A command may be one line or several lines long. A programmer writes commands using a programming language.

see also instruction, operation, statement

a b c d e f g h i j k l m n o p q r s t u v w x y z

A B C D E F G H I J K L M N O P Q R S T U V W X Y Z

command prompt

Command prompt is a way of entering commands into the computer. You type the commands as words.

command word *see* keyword

comment

1 (also reply) An internet comment is a response to online content. Social networks let people interact. One person will post a message or other content, and another person will add a reply or comment.

see also social network

2 A comment is a line in a computer program. It is not a command and is ignored by the computer. Comments are added by programmers as messages to other programmers. They help to explain the code for the human reader.

commonly used password *see* password

comms (*also* communications)

Comms is short for *communications*, a general term for sending messages. Computers communicate by sending digital signals. Comms links let computers share digital data.

see also digital

communication standard *see* protocol

comparison operator *see* operator

compatible

Software or hardware is compatible with a particular computer if it is able to work with that computer. There are different makes of computer. Software made for one computer cannot always run on another type. Hardware made for one computer cannot always plug into another type.

> compatibility

Compatibility means that software or hardware is able to work with a particular computer.

compiler

A compiler is a piece of software that translates source code into machine code. It translates all the instructions in the source code into machine code. It creates an executable file of the machine code instructions.

see also translate, code, executable file, interpreter

complexity

Complexity describes something that has lots of parts, and lots of relations between the parts. It is the opposite of simplicity.

> complex

A complex system has lots of parts, and there are lots of relations between the parts. A complex system or problem is more work for a computer to solve than a simple one.

component

A component is a part of something. A computer system is made of many components. Many of the most important components are electronic.

see also electronic

compress

To compress a file means to make it smaller. Many files are very large and take up a lot of storage space. Compressed files take up less space.

see also zip

> compression

Compression is when a file is compressed.

lossless compression

Lossless compression is when a file is compressed but none of the data in the file is lost.

lossy compression

Lossy compression is when a file is compressed and some of the data is lost, so that the file is lower quality.

computer

A computer is a machine with a processor inside. Computers come in a wide range of shapes and sizes. Other devices can be attached to make a computer system.

see also process

Computer Based Training *see* CAL

computer game *see* game

computer-generated

Computer-generated is a general term to describe anything made by a computer, such as a sound, a voice or a photo.

computerise (*also* computerize)

To computerise something means to give some or all of the work to a computer to do. For example, a computerised quiz is a quiz where the computer asks the questions and marks the answers.

computer-literate

If a person is computer-literate, they can use the computer well. That means two things. Firstly, they do the right actions and can make the computer do what they want. Secondly, they understand the reasons why they need to do these actions.

computer network

A computer network is a group of computers connected by comms links.

see also LAN , WAN, internet

computer science

Computer science is a subject in education. Students of all ages can study this subject. They learn about the science, maths and logic of computers. People who study computer science usually learn to write programs.

computer system

1 A computer system is a computer plus other devices. The other devices might include input, output and storage. Together the computer system has all the devices you need to do a task.

see also computer

2 A computer system is also a group of computers linked together.

A B C D E F G H I J K L M N O P Q R S T U V W X Y Z

computing

Computing is the use of computers. To compute literally means to work out the result of a calculation. Computers are good at calculations, which is why they are called computers, but people also use them to help with many other tasks.

conditional loop *see* loop

conditional statement *see* if statement

conditional structure *see* branching structure

condition-controlled loop *see* loop

configuration

Configuration means how the parts of a computer system are set up so that the system suits your needs.

> configure

To configure a computer system is to set it up in a way that suits your needs.

connection speed *see* bandwidth

connectivity

Connectivity means any way of making a comms link.

> connect

To connect a computer is to join it to the internet or a computer network.

console *see* games console

constant

A constant is a named area of computer memory that stores a value. Once the value is set it cannot be changed, unlike a variable. The value stays constant.

see also variable

contact *see* email

contactless payment

Contactless payment is a way of paying for something using a chip and PIN card. The computer can check your card from a short distance by using radio waves. You do not need to enter the PIN code number.

see also Chip and PIN

content

Content is a general term for words, images, video and music. Anything creative that a person can make is content. Digital content is content made or shared using a computer.

see also social media

contents *see* email

context-sensitive menu *see* menu

control *see* automate

control-click *see* click

control device

A control device is an electrical device that controls something in the real world, such as an electrical engine. By using a control device a computer can control machinery.

control key (*also* Ctrl)

The control key is on the keyboard. It is often marked 'Ctrl'. You usually press the control key at the same time as another key. This is called a keyboard shortcut.

see also keyboard, keyboard shortcut

controller

1 A controller is a part of the processor. It lets the computer operate a control device.

2 A controller is also an input device. It lets the user control the computer, for example when playing a computer game.

cookie (*also* internet cookie, browser cookie)

A cookie is a file with data from a website. A website you visit might put a cookie onto your computer. The cookie stores information about you. Next time you go to the site the cookie tells the website who you are.

coordinates (*also* co-ordinates)

A coordinate is a way of recording the position of a point on a display or on a page. The position is stored as two numbers: x and y.

see also axis

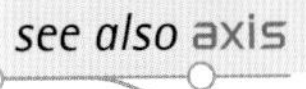

x coordinate

The x coordinate stores the horizontal position of a point. The higher the number the further to the right the point is.

y coordinate

The y coordinate stores the vertical position of a point. The higher the number the further up the screen the point is.

copy

If you copy a file, you make another file with the same contents. You must give the second file a different name or storage location.

copy and paste *see* cut and paste

copyright

Copyright is a law against copying. It is important in computing because computers make copying easy. You do not have the right to make a copy of someone else's work without permission. Nowadays copyright is part of Intellectual Property Rights (IPR).

see also IPR, piracy

corrupt

If data or a file is corrupt, it is spoiled and you cannot read it any more. Hardware and software faults can cause this. If you have made a backup you can restore the file.

see also backup, restore

counter-controlled loop *see* loop

counter variable (*also* loop variable)

In programming, a counter variable is used to count how many times a loop repeats. When the counter reaches the stop value the loop will stop repeating.

see also loop

CPU

The CPU is the part of a computer at the core of the processor, where the computer processes data. CPU stands for *Central Processing Unit*. The CPU is made of the Control Unit and the ALU. The other part of the processor is the memory unit.

see also process, memory unit, ALU

crash

If a program crashes, it stops working. You may need to reboot the computer.

see also reboot

Creative Commons

Creative Commons is the name of a non-profit organisation that issues licenses. This allows people to use digital content that someone else has made. The person who made the file still has IPR.

see also IPR

CSS

CSS stands for *Cascading Style Sheet*. A CSS file sets the style of a web page, for example the colour and layout. It is made by a web designer. The CSS 'cascades' to all the pages on a website, so that the pages all have the same style.

see also HTML

Ctrl *see* control key

Ctrl-Alt-Del *see* keyboard shortcut

curly brackets *see* brackets

cursor

On a computer screen, the cursor is a little vertical line that flashes on and off. It shows the place in a document where your typing will appear. You can move the cursor to a new place in the document by moving the mouse pointer to the place you want and clicking the mouse button.

see also pointer

customised (*also* customized, bespoke)

Customised means adapted to suit a particular user. Customised software has been adapted in order to make it better suited to the needs of one user or organisation.

cut and paste

Cut and paste and copy and paste are common commands that let you move text in a document, images or a complete computer file from one location to another. For example,

- select some text
- use the cut command to delete the text from its original location and put it into the clipboard
- move the cursor to a new place
- use the paste command to put the text from the clipboard back into the document. You can paste many times, so you get many copies of the text. You can use the same process to move or copy other types of objects to new locations.

Cut and paste means the item is moved to the new place. Copy and paste means there are now two versions of the item.

see also clipboard, tool, cursor

WORD BUILD

> copy and paste

The copy command puts a copy into the clipboard, but it does not delete the original text from its position.

> keyboard shortcuts

Cut and paste can be done using keyboard shortcuts: **Ctrl-X** means cut; **Ctrl-C** means copy; **Ctrl-V** means paste.

see also keyboard shortcut

> object

Objects that are not blocks of text, such as images, can also be cut and pasted.

> toolbar

The edit icons on the toolbar can be used to cut and paste.

see also toolbar

cutter

A cutter is a machine controlled by a computer that cuts a substance away in order to produce an object that matches a design. It might cut away plastic or metal, for example.

see also CAD/CAM, 3D printer

cyber-

Cyber- is a prefix that means related to computers. For example, cybercrime means a crime done using a computer.

cyber-bullying

Cyber-bullying is bullying over the internet.

see also social network

⚠ WATCH OUT!

People can say hurtful things on social media. Nobody has to put up with cyber-bullying. Reject online contacts that make you unhappy. In many countries there are organisations that give helpful advice or can be contacted. For example Childline in the UK.

OK

<Dd>

dance mat

A dance mat is an input device used in some computer games. As you dance on the mat the computer detects the movement of your feet.

data

Data means the facts and figures that can be processed by the computer. In programming, a value is one item of data. Expressions state values.

see also value, expression, data type, personal data, data structure

database

A database is a collection of all the data needed for a purpose. It is usually made up of several data tables.

see also DBMS, relational database, data table

database software *see* DBMS

data capture

Data capture means measuring or recording data. It might also include inputting that data. Computers often use automated data capture.

see also input

A B C D E F G H I J K L M N O P Q R S T U V W X Y Z

data check *see* input check

data logger

A data logger is a device that records data for input to a computer. For example, it might record traffic on a road, or the temperature.

> data logging

Data logging is recording data for input to a computer.

data processing (*also* information processing)

Data processing means taking data and organising it into a more useful form, which we call information. The work of a computer could be described as data processing.

see also information, data

data protection

Data protection is a set of laws about using personal data. Most countries have data protection laws. You are not allowed to get, keep or use anyone's personal data without permission. Data must be stored securely.

see also personal data, sensitive data

data structure

A data structure is a named group of memory locations that are linked together. In a linear data structure the memory locations are joined into a line. The different locations are called elements.

see also location

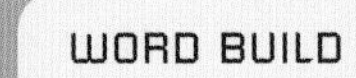

WORD BUILD

> array

An array is a structure that holds a block of data. It can be a line of data values, or a table, or an even bigger structure. It is fixed in size, and does not shrink or grow. The memory locations can store values or they can be empty.

> element

The memory locations of a data structure are called elements. Each element has a name. In a linear data structure, each element is called by the name of the data structure, plus a number. The number is called the index number. In many programming languages the index number is shown in square brackets after the name, like this: `array_name[3]`

> list (*also* linked list)

A list is a linear data structure. It is not fixed in size, and can grow or shrink as values are added or taken away. It has no empty locations.

data table

A database has one or more data tables in it. Each data table has all the data about one type of thing, such as students in a school database.

see also database

data type

The computer uses different methods to store data. These are called the different data types.

see also value, expression

WORD BUILD

> Boolean

Boolean values are True and False. These values are sometimes shown as 1 and 0. There are no other Boolean values.

> float (*also* floating point, real, real number)

Float data type includes all number values. We sometimes call them 'real' numbers or 'floating point' numbers. The numbers can include a decimal point. For example, the number 12.5 is a floating point number. Floating point numbers can be minus numbers.

> integer

Integers are whole numbers and do not include a decimal point. Integers can be minus numbers.

> string (*also* text, character)

String values are made of text characters, such as the text characters you can type with a keyboard. String values are usually shown in quotation marks: "hello world" is an example of string data. String data cannot be used in calculations.

data type conversion (*also* data type casting)

Data can be changed from one data type to another, which will change how the computer stores the data. A string variable might be changed to an integer or a number changed to a string. For example, the numerical value 5 could be changed to the string '5'. Remember a string like this cannot be used in calculations. This is called data type conversion or data type casting. In programming, data type conversion is typically done by a predefined function.

date format

Date format is a way of storing the date in a structured way. The day, month and year are stored as three numbers. In the UK the day comes first, so 03/05/19 is the 3rd of May 2019. In the USA the month comes first, so 03/05/19 is the 5th of March 2019.

date picker

A date picker is a feature of online forms. It typically shows you a list of dates and often looks like a calendar. You pick the date you want by clicking on it.

2019
Thurs, November 07

← NOVEMBER 2019 →

S	M	T	W	T	F	S
27	28	29	30	31	01	02
03	04	05	06	07	08	09
10	11	12	13	14	15	16
17	18	19	20	21	22	23
24	25	26	27	28	29	30

DBMS *also* database software

DBMS stands for *database management system*. DBMS software is an application that lets you make a database. It also lets you get data from the database.

see also database

DDOS *see* denial-of-service attack

dead link *see* broken link

debug

To debug a program means to find and fix all the errors in it.

see also bug, error

A B C D E F G H I J K L M N O P Q R S T U V W X Y Z

declare

In programming, declaring means making something and giving it a name. You can declare such things as procedures, functions and variables.

› declare a subroutine

To declare a subroutine means to make the subroutine and give it a name. This is also called defining the subroutine.

see also subroutine, procedure, function

› declare a subroutine

To declare a variable means to make a variable. This procedure sets aside an area in memory and gives it a name. In some programming languages you must declare a variable before you can assign a value.

see also assign

decomposition

Decomposition means breaking down a big problem into smaller problems. This makes the problem easier to solve. Decomposition is part of planning a program.

decompress

If you decompress a compressed file, you turn it back into an ordinary file.

see also compress

decrypt

If you decrypt an encrypted file, you turn it back to an ordinary file. A lot of internet traffic is encrypted and decrypted. The processor of your computer handles encryption and decryption automatically.

see also encrypt

default

A default is a standard setting or a standard value that is set in advance. If you do not change settings, the computer will use the default value or setting.

defrag (*also* defragment)

To defrag a computer means to remove fragmentation. The computer reorganises files so they are not fragmented, which can make the computer work faster.

see also fragment

› defragmentation

Defragmentation is the reorganisation of computer files so that they are not fragmented.

Del *see* delete key

delete

To delete means to clear or remove something. For example, you can delete a word from a document or delete a file from a folder. Deleted files are stored in the recycle bin.

see also recycle bin, restore, undo

> ⚠ **WATCH OUT!**
>
> If you delete something by mistake you can get it back. You can restore a file from the recycle bin. Other deleted content can be brought back by 'undoing' the delete. If you delete it from the recycle bin however it is really gone and cannot be brought back.
>
> OK

delete key (*also* Del)

The delete key is on the keyboard. It is often marked 'Del'. The delete key deletes whatever is already selected. For example, if you have selected an image in a document, the delete key will delete it.

see also backspace key, keyboard

denial-of-service attack (*also* DOS attack, DDOS, distributed denial of service)

A denial-of-service (DOS) attack is a way to harm an organisation by shutting down its computer system. The computers are flooded with messages and signals and get overloaded. A common method is a DDOS (distributed denial-of-service) attack. That means a group of computers work together to send the messages, making it even more difficult to block them.

desktop

The desktop of a computer is the display you can see behind any open windows on the screen. It is what you see on the screen before you have opened any other windows. You can save files or apps to the desktop. You will see them as icons on the desktop. The recycle bin appears as an icon on the desktop.

see also icon, background, window

WORD BUILD

> taskbar

The taskbar is part of the desktop, usually at the bottom of the screen. It shows the applications that are open. They are shown as icons. By clicking on an icon you can bring that window to the front.

see also icon

> wallpaper

The wallpaper is a picture or pattern that appears on the desktop display. You can choose any picture you like. The wallpaper is sometimes called the background or desktop background.

desktop computer

A desktop computer is a computer that can only fit on the top of a table or desk. It is not portable. The attached devices such as screens can be quite large. This can make it easier to use (but harder to move!).

desktop publishing (*also* DTP)

Desktop publishing is a type of software that lets you make documents. It gives you a lot of control over the layout of each page. Nowadays ordinary word-processing software offers many of these features.

see also word processor

device

Device is a general term for a small piece of equipment. A computer system always has input and output devices.

see also input device, output device

A B C D E F G H I J K L M N O P Q R S T U V W X Y Z

dialogue box (*also* dialog box)

A dialogue box is a small window that opens on the screen when you have to make a choice. The dialogue box often shows you a message or asks a question, and usually has buttons showing choices. You read the choices and click on the button that is right for you.

see also button

Do you want to save the changes made to the document "Document1"?

Your changes will be lost if you don't save them.

Don't Save | Cancel | Save...

dial-up

Dial-up means using a telephone line (a land line) to join a computer to the internet. This is usually a slower type of connection. When you want to use the internet, the computer sends a signal to open the phone line. When you stop using the internet the connection is broken. While you are using the internet your phone cannot be used for ordinary phone calls.

see also broadband

digit

A digit is a symbol that represents a number value. Each digit stands for a different number value. We make numbers by putting digits in the right order. The position of each digit in the number gives the total value of the whole number.

see also number base

digital

Digital means 'made of digits'. Digits are number symbols. Data inside the computer is stored in digital form, which means that it is made of 1s and 0s. All digital data must be stored in number form. However, some real-life data does not exist in exact number form. Data which is not in digital form is called analogue.

see also analogue

digital camera

A digital camera is a type of camera that takes photos that are stored as digital files.

digital device

A digital device is a device that uses digital data. Most computer devices are digital.

directory *see* folder

disable

To disable software means to stop it from working.

disk (*also* disc, diskette)

A disk is a flat, circular piece of plastic or metal that can be used to store digital data. A CD is a plastic disk that stores data using optical storage. A hard disk is usually made of several metal disks joined together.

see also hard disk, optical storage

display

The display is whatever appears on the computer screen.

distributed denial of service *see* denial-of-service attack

distributed processing

Distributed processing is when the work of processing is shared between lots of computers. By sharing the work it is finished quicker.

see also process

div (*also* integer division)

Div is an arithmetic operator. It is sometimes shown by the word DIV or the symbol //. Div gives the result of a division as a whole number (integer).

It leaves out any remainder or fraction. The mathematical name for this is integer division.

see also arithmetic operator, MOD

DNS

DNS stands for *Domain Name System*. This is the system that gives every web page its own address.

see also TLD, URL

docking station (*also* dock)

A docking station is a holder for a portable computer. It lets you charge the battery. It also connects the computer to input and output devices, such as a computer screen or mouse. It makes a laptop or tablet as easy to use as a desktop computer.

see also 2-in-1 device

document (*also* document file)

A document is an object made of text. A document file stores the text of a document. Most document files also store the format of the document, and any images.

see also file extension, object, text

WORD BUILD

> document design

The design of a document is how the document looks, including text format, alignment and so on. Many word processing applications let you choose a design for your document.

see also format

domain

Locations on the internet are organised into domains. There are large domains called Top Level Domains (TLDs). Within the TLDs are smaller domains controlled by individual organisations.

see also DNS, domain name, TLD

domain name

A domain name is a string of characters that identifies a location on the internet. Here is an example: www.bbc.co.uk This domain name has several parts:

- .uk is a Top Level Domain (TLD). This identifies that the domain is registered in the UK.
- .co is a second level domain. This identifies that the domain is a company within the UK.
- bbc is a lower level domain. This identifies the particular company that controls the location.
- www stands for world wide web. This identifies that the location is a web server controlled by the BBC.

dongle

A dongle is a small device that plugs into a socket on a computer. You do not have to open the computer case to plug it in. The dongle lets your computer do new things. For example, a Bluetooth dongle will let your computer use a Bluetooth link.

see also Bluetooth, plug-in, Wi-Fi

DOS attack *see* denial-of-service attack

dot-com *see page 6*

double-click *see* click

download

1 To download a file, such as music or a video, means to get a copy of it through an internet connection. After you have downloaded the file, you have a copy of it on your computer.

see also upload

2 A download is a copy of a file that you download to your computer through an internet connection.

a b c d e f g h i j k l m n o p q r s t u v w x y

A B C D E F G H I J K L M N O P Q R S T U V W X Y Z

DPI

DPI stands for *dots per inch*. This is a measure of print quality. Lots of small dots make the image clear and detailed. Inkjet printers use 300-700 dots per inch. Laser printers can use more than 2000 dots per inch.

see also resolution, printing

draft

A draft is an early version of something. It is a good idea to start with a draft. You might make a draft document. You can change the draft later, for example by correcting mistakes or adding extra features.

drag

Dragging the mouse means moving the mouse while holding the mouse button down. You can click on an object, and leave the button pressed down. As you drag the mouse, the object moves with it.

see also mouse

drag and drop

When you are moving an object, you can move the object to the right position, then let go of the mouse button. The object will stay in the new place. This action is called drag and drop.

drive

1 A drive is the hardware that connects the computer to storage. For example, a CD drive lets the computer use CD storage.

see also optical storage

2 A drive is also the computer's built-in storage, such as the hard disk or SSD.

see also hard disk, SSD

3 Every file has a location, which is where the file is stored. The file location may be shown using a capital letter followed by a colon. The letter stands for the drive. For example a file location on the hard drive in Windows begins C:\.

see also file location

driver

A driver is a small piece of software that lets the computer use an item of hardware. Each make of hardware has its own driver. When you plug in a new piece of hardware, such as a new mouse or headphones, the computer will find and load the right driver. Most drivers you might need are already installed on your computer. You can download others.

see also plug and play

DRM

DRM stands for *digital rights management*. DRM is the use of technology to protect copyright (or IPR).

see also copyright, IPR

drone (*also* UAV)

A drone is a small flying robot. Some are remote-controlled by humans on the ground, while others are controlled by a computer inside the drone. Another name for a drone is UAV (Unmanned Aerial Vehicle).

DTP *see* desktop publishing

dumb terminal *see* terminal

DVD *see* optical storage

DVI

DVI stands for *Digital Video Interface*. DVI is a type of port on some screens that is used to connect a video device to the computer.

see also port

dynamic content

Dynamic content means content which changes or moves, such as animation or a web page.

see also content

earphones (*also* earbuds, ear-fitting headphones)

Earphones are a type of headphone. An earphone or earbud is the right size and shape to fit inside the outer part of your ear. This holds the earbud in place.

see also headphones

Easter egg

An Easter egg is a hidden feature in the form of a joke or message for users who know how to find it. Easter eggs can be found in lots of digital products, for example games, videos and software. For example, if you hold an Apple Mac wired mouse above your desk, the red light shining from the bottom of the mouse will make the shape of a mouse's head. Hidden features like this can be fun for users. They may pass news of the secret feature to their friends. This is a form of viral marketing.

see also game, viral marketing

eBay

eBay is the name of a business that runs a large website that helps people to buy and sell things. The people who sell can be businesses or individuals. Sometimes people sell by auction.

see also e-commerce

e-book

An e-book is a book in electronic form. It can be read by an e-reader such as a Kindle.

see also e-reader

e-commerce

E-commerce means buying and selling things using the Internet.

see also Amazon, eBay

Edge

Edge is the name of a web browser made by Microsoft.

edit

To edit means to make changes to text or data. Computers make it easy to edit information, images and documents.

edit tools *see* tool

electronic

1 Electronic describes a device that can control or change the flow of electricity that passes through its circuits. A computer processor is an electronic device.

see also process

2 Electronic also describes things that people do using a computer or the internet, such as electronic mail (email) or electronic banking.

see also processor

electronic gate *see* gate

element *see* data structure

A B C D E F G H I J K L M N O P Q R S T U V W X Y Z

else *see* if statement

elseif (*also* elif) *see* if statement

email

Email is a way to send and receive messages using an internet connection. The messages are in text but you can attach files with other content such as images or sounds.

email address

An email address tells the computer where to send an email. An email address always includes the symbol @. The part after the @ shows the domain that will receive the email. The part before the @ shows the person who will get the email from the server.

see also IP address, memory address, web address

WORD BUILD

actions

When you receive an email you **open** it to read it. You can then: **forward** it to someone else; **reply** to the person who sent it; **reply all**, which means that the email goes to everyone who got the original email.

contact

An email contact is a person and their email address. You can store the contacts you know in a contact list. This is like an address book for emails.

contents

When you receive an email you will see: the contact details of the person who sent it; the contact details of other people who got the email; the date and time it was sent; a subject line telling you what the email is about; the body of the email, i.e. the main text; any attachments.

folders

Email software usually has several folders. They could include: **Inbox** (holds emails you have received); **Outbox** (holds emails ready to be sent); **Sent items** (holds emails already sent).

send

When you send an email you must give the email address to send it to. There can be more than one address. You can also use **cc** (so that other people get a copy of the email) or **bcc** (so that people get copies of the email, but their addresses are not shared).

email server *see* server

embedded video

An embedded video is a video that you can see inside a web page.

emoji

An emoji is a little picture included in an internet message or text message. You add it to show your feelings or for fun. The word comes from the Japanese for picture. Every emoji has a character code.

see also ASCII, Unicode

emoticon

An emoticon is a little picture included in an internet message. It is made of other characters such as punctuation marks. An example of an emoticon is :-)

emulator

An emulator is a piece of hardware or software that makes a computer act like a different type of computer. For example, it might make your desktop computer show a screen display like a phone.

encode

To encode data is to put it into code. This is not always a secret code. For example, all data must be encoded as binary numbers when it is input to the computer.

see also encrypt, input

encrypt

To encrypt something means to put it into a secret code. Data is often encrypted before it is sent over the internet. Your computer does this automatically.

see also decrypt

> encryption

Encryption means putting something such as data into a secret code.

enter

To enter means to input into the computer. The user can enter a command or enter some data.

see also input

enter key (*also* return key)

The enter key is a large key on the keyboard to the right of the letters. When you have entered some data or a command you press the enter key in order to complete the action. In a word-processed document the enter key is used to make a new line. It is also called the return key.

see also keyboard

EPOS (*also* POS)

EPOS stands for *electronic point of sale* often just referred to as the *point of sale*. It is a computerised system used in shops to pay for goods. It is sometimes called a till.

equals sign (*also* equal sign)

The equals sign is the symbol =. This symbol is used as an operator in many programming languages. As a relational operator it is used to check whether two values are the same. As an assignment operator it is used to assign a value to a variable. In most programming languages the equals sign is used for only one of these purposes. Another symbol is used for the other purpose.

see also assign, operator

e-reader

An e-reader is a standalone device that lets you download and read e-books such as a Kindle.

see also e-book

error

An error is a mistake. An error in a program stops it from working the way the programmer wants it to.

logical error (*also* logic error, semantic error)

If a program has a logical error it does the wrong thing. This may be because the programmer has written the wrong instructions. For example, the program might tell the computer to multiply instead of divide. The program runs and there is no error message, but the result will be wrong. The programmer must test the program carefully to make sure there are no logical errors.

runtime error (*also* run-time error)

If a program has a runtime error the program includes commands that are impossible for the computer to carry out. An example is example to divide by 0. Another example is a loop that does not have an exit condition. If there is a runtime error the program will crash or get stuck. There may be an error message to tell you what went wrong.

syntax error

Syntax means the rules of a language. Programming languages have syntax. A syntax error means the program breaks the rules of the programming language. If a program has a syntax error it cannot be translated into machine code. You cannot run the program.

user error (*also* input error)

A user error means the user has entered the wrong data. For example, they might type a letter instead of a number. This can make the program go wrong. Good programs have validation checks to catch user errors.

see also validation

error message

If the computer finds an error it stops the program and shows an error message. The error message helps the programmer to find and fix the error.

escape key

(*also* Esc)

The escape key is on the keyboard, usually at the top left. It is often marked 'Esc'. The escape key is used to escape from (exit or cancel) a task. The task stops and work will not be saved. The escape key does not work with all apps.

see also keyboard

eSports

eSports are organised computer gaming competitions. People gather to watch expert teams compete against each other in computer games. They also watch over the internet. It is estimated that about 500 million people watch eSports every year. Professional players make a living by playing video games.

Ethernet

Ethernet is a way to join computers to a network using cable.

event-driven programming

Event-driven programming is a way of writing programs. The commands in the program are linked to events. If the event happens, the commands are carried out. If the event does not happen, the commands are not carried out.

see also trigger

Excel *see* MS Office

executable code

Executable code means instructions the computer can carry out. Machine code is executable code.

see also code, execute

executable file

An executable file is made of machine code instructions. It has the file extension .exe. When you run the file, the computer will execute (carry out) all the instructions in the file.

see also code, file extension, execute, run

execute

Execute means carry out. To execute an instruction means to carry out the instruction. To execute a command means to run the command.

see also instruction, run

exit

To exit a task or app means to stop it, for example by clicking the close button.

see also close button

exit condition

In programming, an exit condition is the method used to stop a loop. Some loops are stopped by a counter. Others are stopped by a logical test. If a loop does not have an exit condition it will repeat forever.

see also loop

expansion card

An expansion card is a circuit board that gives the computer extra features or extra power. You can plug it into your computer. It fits into an expansion slot so you need to open the computer case first.

see also circuit board, expansion slot

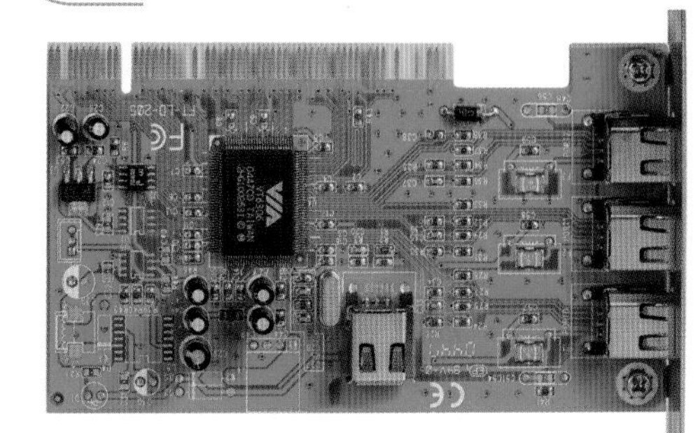

expansion slot

An expansion slot is a space inside a computer case that you can put expansion cards into.

see also expansion card

expert system

An expert system is software that makes decisions. It follows rules to get information and choose a good decision.

export

To export content means to send it from one file into another file, often in a different format.

see also import

expression

In programming, an expression is a way of stating a value. Many program commands include an expression. There are many types of expression. A Boolean expression holds the values True or False. A numerical expression holds a number value. A text or string expression holds a series of keyboard characters. An expression may include an operator. For example, the expression 12 + 5 expresses the value 17.

see also Boolean, operator, value

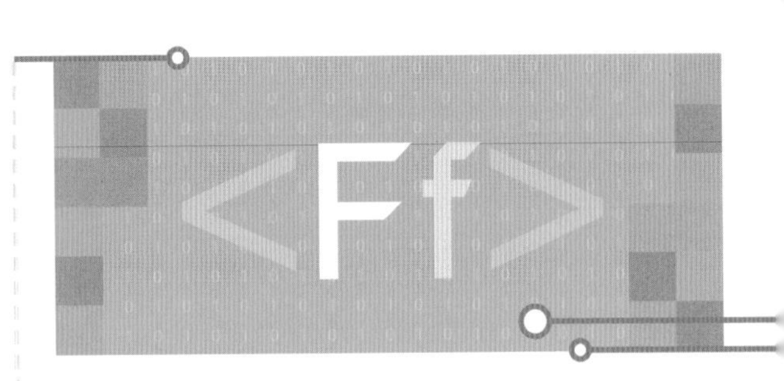

extension *see* file extension

external memory

External memory is a kind of storage used for storing data outside the processor.

see also storage

extranet *see* intranet

eye scan *see* biometrics

eye strain

Eye strain means eye problems caused by too much time looking at a computer screen.

> ⚠ **WATCH OUT!**
>
> Take regular breaks from the computer to avoid eye strain.
>
> OK

e-zine

An e-zine is a magazine you read on the internet.

F2F

F2F is short for *face to face*. It is used to talk about meeting a person in real life instead of on the internet.

see also IRL

Facebook

Facebook is the name of a company that runs a social network. It is the largest social network in the world.

see also social network

Facetime

Facetime is a system for video chat messages. You can get a Facetime app and put it on your phone or other device. Facetime allows one-to-one or group chats.

see also chat

facial recognition *see* biometrics

A B Q R S T U V W X Y Z

factory settings (*also* factory preset)

Computers can have different settings which change how the computer behaves. The factory settings are the ones set when the computer is made, so when you get a new computer it uses the factory settings. Factory settings are default hardware settings.

see *also* default

FAQ (*also* FAQs)

FAQ stands for *Frequently Asked Questions*. This is a list of common questions about a certain subject or product, together with their answers.

see *also* help

WATCH OUT!

You may see FAQs when you look for help with a problem. Read through the FAQs to see if your problem has been answered already.

OK

favourite *see* bookmark, like

feedback

Feedback is when the results of a process affect the process itself. This general idea is used in many different ways in computing. For example, if people tell you what they think about something you made, the feedback might help you improve it. Another meaning of feedback is when the computer measures the system that it controls. The feedback helps the computer to adjust its settings.

feedback loop

A feedback loop is how the computer controls a process or system. A control system has an input, a process and an output. Output from the computer affects a system. Input to the computer provides information on the same system. The computer uses the inputs to adjust its outputs.

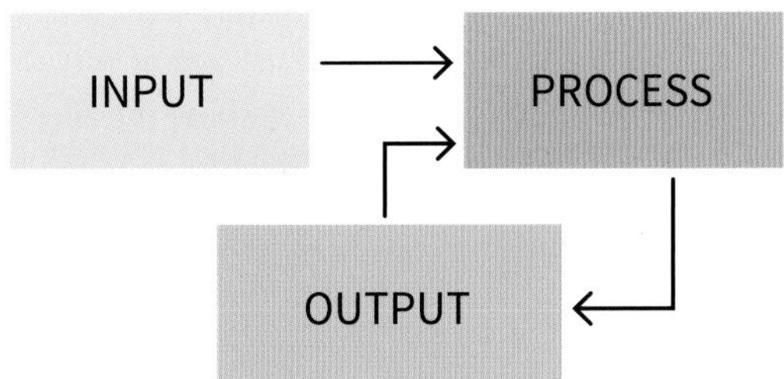

fetch-execute cycle (*also* instruction cycle)

The fetch-execute cycle describes the action of a computer processor. Instructions are *fetched* from memory and then *executed* (carried out).

see *also* process, instruction, execute

fibre-optic (*also* optical fibre)

Fibre-optic cables are made of fibre glass. They can be used to send digital signals in the form of pulses of light.

field

In a database, a field stores one item of data. Each column of a data table stores one field.

see *also* data table

file

The contents of computer storage are organised into files. There are two types. Software files are executable files which you can open and run. Data files are not executable and are opened within other software applications.

A B C D E F G H I J K L M N O P Q R S T U V W X Y Z

file extension

Every filename has an extension. The extension is usually three letters at the end of the filename, placed after a dot. It tells you what type of file it is. Some computers are set up to show the file extension, while others do not show it.

see also image file

file icon

A file is shown on the screen as an icon. The type of icon usually tells you what type of file it is.

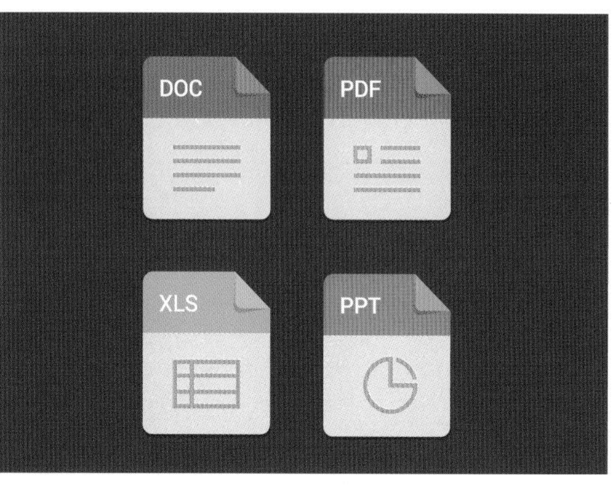

file location (*also* file path)

File location tells you where a file is stored. The file location starts with a letter, for example C:\ which tells you the drive. It then goes on to show the folder and sub-folders where the file is stored.

see also drive, folder

file manager

The file manager is a feature of an operating system. It lets you see the files that are stored on your hard drive or available over a network. You can **open**, **copy**, **move** or **rename** files. The files are usually shown as icons.

see also operating system

file menu (*also* file tab)

Many apps have a file menu or file tab. The file menu lets you save and open files. It also lets you start a new blank file.

filename

Every file has a unique name. The name is made of letters, numbers and sometimes other characters. The name of the file should remind you of its contents.

file sharing

1 File sharing can mean distributing a file to other users. Each user ends up with their own copy of the file. There are websites that share music files, games etc. It can be illegal to share a file without permission.

see also copyright, streaming

2 File sharing can mean that several people have access to the same file. There is one copy of the file, but several people can see it and edit it. Often the file is stored in cloud storage. This type of file sharing helps with team tasks and collaboration. Web services like Google docs and Dropbox allow this type of file sharing.

see also cloud-based

file size

File size is usually measured in bytes. It is shown as a number followed by the letter B. B stands for *Byte*. Sometimes a file size is measured in bits, in which case a lower case b is used instead. As file sizes can be very large, we often use orders of magnitude (such as megabyte) to simplify the number.

see also bit, byte, order of magnitude

file type

Different types of file can do different things. You can tell the file type from the file extension or from the image used as a file icon.

filter

In a database, a filter is a way to see only some of the records in a data table. A filter is usually based on one field of the table. You can choose the records you want to see, for example all students taking a particular subject.

see also data table, record

find

Find is a command that allows you to look for a match. For example you could tell the computer to find a particular word in a document file. You click on the find tool, then type the text you want to find. The computer will find all matching text. The keyboard shortcut for find is Ctrl-F.

> find and replace

Find and replace is a command that allows you to find all matching text in a document and then swap it for different text.

finger print *see* biometrics

finite loop *see* loop

Firefox (*also* Mozilla Firefox)

Firefox is a the name of a web browser made by the Mozilla Corporation. You can download it and use it for free.

firewall

A firewall is a barrier between a LAN and the internet. It can be made of software or hardware or both. A firewall checks all the signals that go in either direction and prevents malware getting through.

see also LAN, malware

firmware

Firmware is software that is fixed in the hardware of the computer and is hard to change. We sometimes say the instructions contained in this software are 'hardwired'.

see also software, memory, non-volatile

first-person shooter (*also* FPS)

A first-person shooter is a game genre. The game is displayed from the point of view of a person such as a soldier who shoots enemies or monsters.

see also game genre

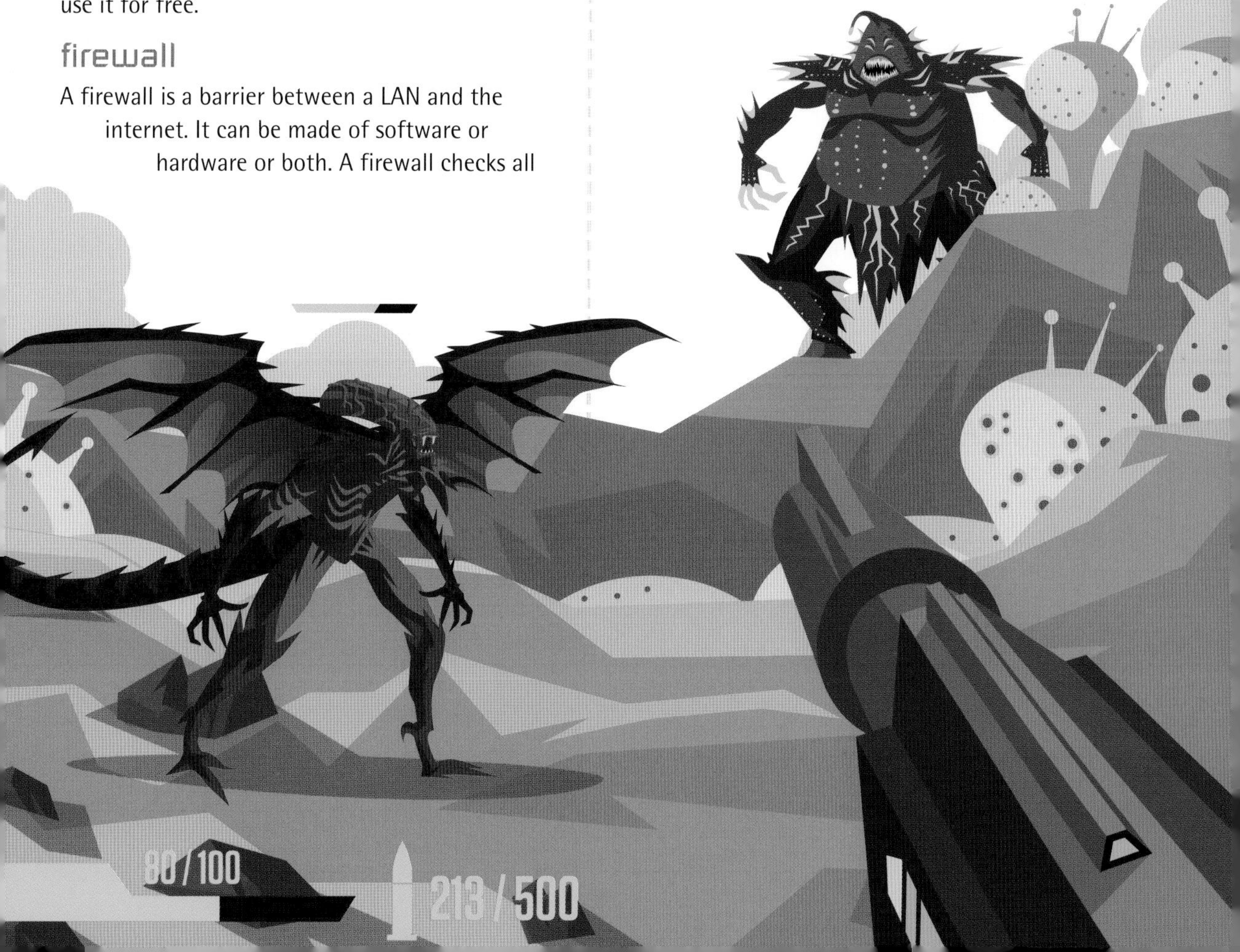

80/100 213/500

A B C D E F G H I J K L M N O P Q R S T U V W X Y Z

fitness tracker

A fitness tracker is a small device that you typically wear like a wrist watch. It logs your exercise and other movements, and then reports how much you have done. A common make of fitness tracker is called a Fitbit.

fixed loop *see* loop

Flash (*also* Adobe Flash)

Flash is a method for making and showing dynamic content such as animations and video. It was made by the Adobe corporation. It is now mostly replaced by more modern methods.

see also dynamic content

flash memory (*also* flash drive)

Flash memory is a way of storing data by using trapped electrons. Flash memory can store a lot of data in a very small space.

see also pen drive

float *see* data type

floating point *see* data type

floor turtle *see* turtle

flops (*also* FLOP)

Flops is short for *floating point operations*. It is a measure of how many operations a computer can do in one second. This tells you how fast the computer is.

see also petaflop, MIPS

flowchart

A flowchart is a way of setting out an algorithm. It sets out the steps to solve a problem, using boxes joined by arrows.

see also algorithm

folder (*also* directory)

A folder is an area containing a number of files, usually grouped together under a common subject. A storage drive might hold thousands of files. Dividing them into folders makes it easier to find the file you want.

see also email

> subfolder (*also* subdirectory)

A folder inside another folder is called a subfolder.

follow

You can follow another user on a social network. This means that you will see their posts.

see also social media

font

Font means the style of text used in a document.

monospace font

In a monospace font all letters are the same width. The most common monospace font is Courier.

Aa

sans serif font

Sans serif fonts have plain letters without small lines (called serifs) at the top and bottom. Arial is a common sans serif font.

serif font

Serif fonts have small lines (called serifs) at the top and bottom of letters. Times New Roman is a common serif font.

WORD BUILD

> font size

Font size is measured in points. Normal text is usually 10-12 points. This text is in 10 points.

footer

A footer is text that appears at the bottom of a document page, for example the page number. Many word-processing applications let you create a footer. It will appear at the bottom of every page of the document.

see also header

foreign key

A foreign key is a key field from one database table that is shown in a different table. This makes a link between the two tables. For example, the ID code for a course might go into the record for a student. That links the student to the course.

see also key field, relational database

forever loop *see* loop

for loop

In many text-based programming languages, the keyword 'for' is used to make a counter-controlled loop.

see also loop

format

Format is extra information that goes with computer content. It tells you how computer content will be displayed or organised.

document format

Document format means the layout of a document, for example how the text is organised on the page.

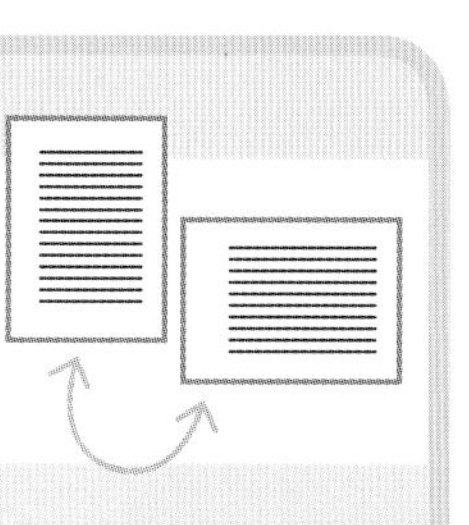

see also page orientation

text format

Text format means the appearance of text, for example whether it is bold or italic.

bold
italic

see also font, bold

formatting *see* show formatting

format tools *see* tool

formula

1 A formula is a type of expression that combines two or more values to give a new value. For example, an arithmetic formula expresses a value using a calculation.

2 In spreadsheets, a formula is a type of instruction that takes values from other parts of the spreadsheet (cells) and performs calculations using them.

see also calculate, spreadsheet, variable

forum (*also* internet forum)

An internet forum is a website where people can post messages and reply. The posts are usually text and stay on the site for some time.

FPS *see* first-person shooter

fragment

To fragment computer files is to store them as lots of small pieces of information in different parts of the computer memory or hard disk.

see also defrag

> fragmentation

Fragmentation is when a file is stored as lots of small pieces of information in different parts of the computer memory. This can happen if memory is very full. Files in storage can also be fragmented.

FPS *see* first-person shooter

A B C D E F G H I J K L M N O P Q R S T U V W X Y Z

freeware

Freeware is software that you are allowed to use for free. For example, many web browsers can be downloaded and used for free.

FTP

FTP stands for *File Transfer Protocol*. It is a standard way of sending a file down an internet connection. Nowadays FTP is used by other internet software apps that are easier to use.

see also protocol

full-screen view

Full-screen view means making the active window so large that you cannot see anything else on the screen.

fully justified *see* aligned

function

In programming, a function is a type of subroutine. A function always creates a new value, which is called the returned value.

see also subroutine, procedure, returned value

functionality

The functionality of a piece of software or hardware is the full range of things that it can do. Software with greater functionality can do more things.

function key

The function keys are the keys along the top of the keyboard. There are 12 function keys numbered from F1 to F12. They are used for different purposes by different apps. F1 often opens a help screen.

see also keyboard shortcut

G *see* order of magnitude

game (*also* computer game, video game)

Computers are used by many people to play games. There are many types of computer game. You can play against the computer or you can play with friends, using the computer to control the game. Some games have puzzles and some have simulated actions and adventures. A computer game is also called a video game.

> gaming

Gaming is playing computer games.

game engine

A game engine is a type of software that helps programmers to make a computer game. Many computer games look the same because they were made using the same game engine.

game genre

A game genre is one of the different types of computer game you can play, such as a sandbox game, an adventure game or a first-person shooter.

game loop
The game loop is part of the software that makes a game work. It is a forever loop (infinite loop). This means the game keeps running the whole time it is switched on.

see also loop

gameplay
Gameplay is whatever happens in a game. Different game genres have different types of gameplay, such as shooting or solving puzzles.

gamer
A gamer is a person who likes to play computer games.

see also game

games console (*also* console)
A games console is a device that lets you play computer games. It usually has a processor in it. Unlike many other computer devices, consoles are not standard and cannot work together.

gaming computer
A gaming computer is a type of computer that is designed to be very good for playing computer games. It might not be so suitable for other uses.

garbage in, garbage out (*also* GIGO)
'Garbage in, garbage out' is a saying used by people who work with computers. It means that if you put incorrect data into a computer then incorrect answers will come out, no matter how good your computer is.

gate (*also* logic gate, electronic gate)
A gate is a microscopic electronic circuit that can switch an electrical charge from ON to OFF. Examples of gates are the AND, OR and NOT gates.

A
B
Z

see also AND gate, OR gate, NOT gate

gateway
A gateway is a link that joins two networks together.

giga (*also* G) *see* order of magnitude

GIMP
GIMP is free software that you can use to make and edit bitmap images. It is based on a type of free software called GNU. GIMP stands for *GNU Image Manipulation Program.*

see also bitmap graphics, freeware

glitch
A glitch is an electronic problem that can make a computer go wrong.

global variable
A global variable is a variable that can be used anywhere in a program.

see also variable, local variable

Gmail *see* Google

a b c d e f g h i j k l m

A B C D E F G H I J K L M N O P Q R S T U V W X Y Z

Google

Google is the name of a search engine made by the company Google. It is the most widely used search engine in the world.

see also search engine

WORD BUILD

> Gmail

Gmail is a service offered by Google. It is a website that lets you send and receive emails. You can see the emails in your browser.

> Google maps

Google maps is a service offered by Google. If you enter a location, the website will display a map of that area.

> Google street view

Google street view is a service offered by Google. If you enter a location, the website will display a photo taken from a camera at that location.

GoPro

GoPro is the name of a company that makes action cameras. These are video cameras used for recording activities such as cycling and climbing. The camera can be fixed to your helmet or bike.

GPS

GPS stands for *Global Positioning System.* GPS is a way to find the exact location of a device using signals from satellites in orbit. It can identify any place on Earth.

see also Sat Nav

graph

A graph is a way to show number information in a visual form. Examples include a pie chart and a line graph. Spreadsheet software can make a graph from a set of numbers.

graphics

Graphics is a general term for computer images. Graphics software is software that lets you create an image.

WORD BUILD

> graphics format

A graphics format is a way of storing an image in the computer. There are two main graphics formats: bitmap images and vector images.

see also bitmap graphics, vector graphics

> graphics object

A graphics object is a shape or area on the screen. Vector graphics are made of smaller objects. Objects can be selected and then cut, copied, resized, stretched and so on.

> graphics tools

Graphics tools are icons in the toolbar of graphics software. Different tools let you do different things such as drawing lines and shapes, and adding colours.

graphics accelerator

see graphics card

graphics card (*also* video card, graphics accelerator)

A graphics card is a type of expansion card that improves the graphics of your computer. A computer with a graphics card can change the image on the screen more quickly. That is good for fast gaming.

see also expansion card

graphics format *see* graphics

graphics object *see* graphics

graphics tools *see* graphics

GUI

GUI stands for *Graphical User Interface*. This is an interface that has graphical features, such as buttons, menus and icons. Almost every app nowadays has a GUI.

see also interface

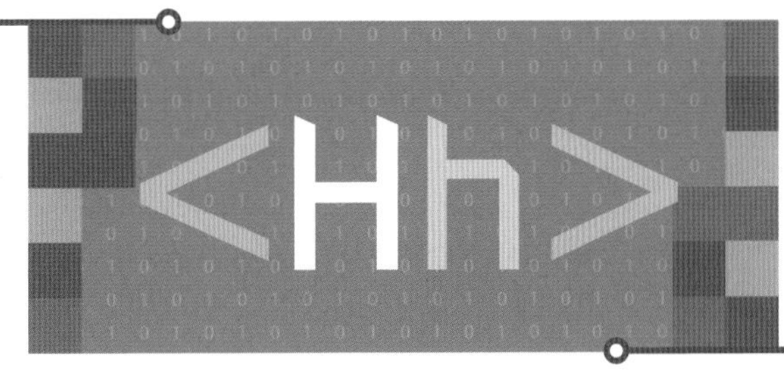

hacking

Hacking means breaking through computer security. A person who does this gets access to private files and data and may steal or destroy data.

› hacker

A hacker is a person who uses a computer to break into a computer system.

hand-held computer

A hand-held computer is a small computer that you can hold in your hand. A tablet is a hand-held computer.

see also tablet, smartphone

handshake

A handshake is a message that two computers exchange before they start sending data. The handshake lets each computer check that the other is authentic.

see also authenticate

hard coded

If data values are included in program code then they are 'hard coded'. To change the values you have to change the program code.

see also hard-wired

hard copy

Hard copy means a printed copy of a computer file. People sometimes call the screen display 'soft copy'.

see also printing

A B C D E F G H I J K L M N O P Q R S T U V W X Y Z

hard disk (*also* hard drive)

A hard disk is a set of metal disks held inside the computer case. It uses magnetic storage to hold data. Nowadays some hard disks use flash memory. The hard disk is also known as the hard drive.

see also magnetic storage

hardware

Hardware is all the devices that make up a computer system.

see also software

hard-wired

If something is hard-wired, it is fixed in a computer system. It is part of the hardware and cannot be changed.

see also firmware

Hashtag *see*

HDMI

HDMI stands for *High-Definition Multimedia Interface*. HDMI is a way of connecting a display to a computer. It can transmit sounds and images.

see also port

header

A header is text that appears at the top of a page. It is not part of the main document. For example, a page header may show the date or the page number. Many applications let you create a header. This includes word-processing or spreadsheet software. When the document is printed the header is added at the top of every page.

see also footer

heading

A heading is a title or subtitle and is usually shown in large or bold text. Using headings helps you to write documents in a structured way. In a word-processed document, headings can mark different sections. In a spreadsheet, headings can be added at the top of column.

see also document

WORD BUILD

> format

Word-processing software can be used to make text into a heading. You select the text, then click on a heading style. The largest and most significant style is usually called 'Heading 1'.

> subheading

A subheading is a smaller heading. A document is often structured with headings and subheadings.

headphones

Headphones are an output device that let you listen to sound output from a computer. They have two very small speakers. The shape of the headphones holds the speakers close to each ear, so that you can hear the sound, but nobody else can.

⚠ WATCH OUT!

If you listen to music too loud on headphones you can damage your hearing. If other people can hear the music from your headphones then it is too loud.

OK

heat sink

All processors produce some heat, but too much heat can make the computer stop working properly. A heat sink is a device that takes heat away from the processor.

help

Help is any feature that helps you use a program if you have a problem with a task. Some apps have a help option on the menu bar or there may be a button marked with a question mark.

help desk

A help desk is a service that provides help and advice. A help desk can be provided for computer users in a large organisation, or it might be provided by a software or hardware company for their customers. Help might be available by phone, by email, or on a website. A key feature is that your problem is handled by a real person.

see also help

Hertz

Hertz means 'number of cycles per second'. It is pronounced like the word 'hurts'. Hertz is one way to measure the speed of a computer. The fetch-execute cycle repeats millions of times per second and the hertz tells you how many times.

see also MIPS, flops, fetch-execute cycle

hexadecimal (*also* hex)

The term hexadecimal is used to describe numbers made in base 16.

see also number base, binary

hierarchy

A hierarchy is a way of organising things. A hierarchy has a small number of items at the top and a larger number of items underneath, rather like a family tree. Folders and subfolders in computer storage are organised into a hierarchy.

> hierarchical

Hierarchical means organised into a hierarchy.

high-level language

A high-level language is a type of programming language. High-level languages are designed to help a programmer to write code. There are many thousands of different high-level languages.

see also programming language, low-level language

highlight

1 To highlight part of a document is to make it stand out, for example by adding a background colour.

2 If you highlight an area of text, you mark it out on the screen to select it before copying it, deleting it, moving it or changing it.

see also select

high-resolution (*also* high-res, hi-res, high-definition)

A high-resolution image has lots of small pixels. This makes the image detailed, sharp and realistic, but it also makes the file size large.

see also resolution

hit *see* page hit

home computer

A home computer is a computer that you use at home, usually a desktop. A home computer may be less powerful than ones found in schools or offices.

see also desktop

home page

The home page is the main page of a website. You can get to the other pages on the site using links on the home page. These pages usually have a link back to the home page.

hot link (*also* hotlink) *see* hyperlink

hotspot (*also* wireless hotspot)

A hotspot is a place where there is a wireless signal. If you are in a hotspot your device can connect to a wireless network.

see also Wi-Fi, WAP

HTML

HTML stands for *HyperText Markup Language*. HTML is the language used to make web pages. When you look at a web page the HTML comes onto your computer. Your browser displays a web page based on HTML.

see also web page, browser, hypertext

HTTP (*also* http)

HTTP stands for *HyperText Transmission Protocol*. HTTP is a method for sending hypertext, such as HTML. If you look at a web page address, you will see that it begins with http. This shows that it is sent to your computer using HTTP.

see also hypertext, protocol, web page

WORD BUILD

> HTTPS (*also* https)

HTTPS is a variant of HTTP with added security. It stands for *HyperText Transmission Protocol Secure*. If a web page address begins https, it is secure and has been checked as a valid website.

⚠ WATCH OUT!

Never send information about your money or personal details to a website UNLESS it begins with https.

OK

hub *see* network

hyperlink (*also* hot link)

A hyperlink is a place in a document that links to another document. If you click on a hyperlink, your computer will open the new document. Nowadays hyperlinks are used to move from one web page to another.

see also browse

www.youtube.com

hypertext

Hypertext is text that has hyperlinks in it.

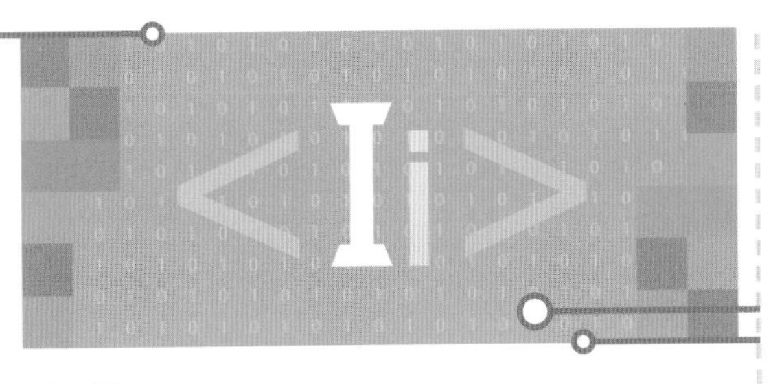

IAS

IAS stands for *Immediate Access Store*. It is another name for the memory unit.

see also memory unit

icon

An icon is a little picture that stands for an app, file or task. You click on the icon to choose the one you want to use.

ICT

ICT stands for *Information and Communications Technology*. At one time people described the use of computers as IT (Information Technology). Once comms links between computers became so important, 'Communications' was added to the name.

see also IT

IDE

IDE stands for *Integrated Development Environment*. An IDE is a software application that has all the features you need to write a program. It lets you enter and edit code, has tools to help you debug the code and lets you run the code. Most IDEs are designed to work with only one or two programming languages.

see also code, run, debug, edit, IDE

identifier

An identifier is a name. In programming, many items are given identifiers, for example subroutines and variables. A good programmer chooses identifiers carefully. The name should remind you of what it identifies. Most programming languages have rules about identifiers. For example in many languages an identifier must be one word only, and start with a letter.

see also subroutine, variable

identity theft

Identity theft means pretending to be someone else, usually by using the internet. It is done by logging in as them and then stealing their money and seeing their data.

see also log in

IDLE

IDLE is an example of an IDE. It is one of the IDEs developed to use with the Python programming language.

see also high-level language, programming language, IDE

Python 3.7.1 Shell

File Edit Shell Debug Options Window Help

```
Python 3.7.1 (v3.7.1:260ec2c36a, Oct 20 2018, 14:57:15) [M
SC v.1915 64 bit (AMD64)] on win32
Type "help", "copyright", "credits" or "license()" for mor
e information.
>>> 2+4
6
>>> "hello"
'hello'
>>> |
```

Ln: 7 Col: 4

if statement (*also* if-then statement, conditional statement)

In programming, an if statement is a statement that starts with the word 'if' and a logical test. The commands that follow are only carried out if the test is True. An if statement can be expanded by adding else or elseif, or both.

see also logical test, program structure

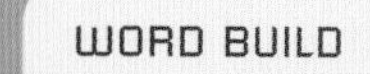

WORD BUILD

> else

The word 'else' can come at the end of an if statement. The if statement includes a logical test. The commands that follow 'else' are carried out if the test is False.

> elseif (*also* elif)

Elseif is a way to expand an if statement. The word 'elseif' is followed by a new logical test. There can be lots of elseif tests. The computer looks down the list of tests until it finds one that is true. It carries out those commands.

iMac

An iMac is a type of desktop computer made by Apple.

see also desktop, Apple, iOS

image

An image is a general term for a picture of any kind, especially one stored in electronic form.

image file

An image file is a computer file that stores an image. Images are stored using a number code system. The different ways of storing images are called graphics formats.

see also graphics

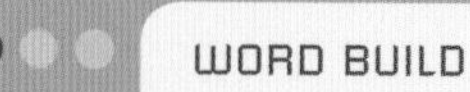

WORD BUILD

> image file extension

All filenames have extensions. The extension of an image filename tells you what graphics format it uses. For example, this is what these filename endings mean:

.bmp: a bitmap file
.jpg (pronounced 'jay peg'): a type of compressed bitmap file
.gif (pronounced 'gif' or 'jif'): another type of compressed bitmap file.

see also file extension

image recognition software

Image recognition software can scan a picture and identify what it is, for example that it is a face or a car.

import

To import content means to bring it into your file from another file.

see also export

inbox

An inbox is an area on your computer that stores the email messages that come in.

incremental backup

Backup is important, but it can take a long time to back up all your files. Typically, a business will make a full backup at regular intervals, for example once a week. In between these times they will make an incremental backup. The computer finds the files that have changed since the full backup. These files are backed up, and no others. This is quicker than making a full backup.

see also backup

indent

To indent text is to set it in from the margin. You can do this in all types of document. Indenting text can make it stand out in the page. In some programming languages, indenting is used to mark a block of code, such as the commands inside a loop.

> indentation

Indentation is setting text in from the margin.

index number *see* data structure

infinite loop *see* loop

information

Information is data that has been organised or processed to make it more useful. Computers are used to turn data into information.

see also data processing, data

information processing *see* data processing

initialise (*also* initialize)

Initialising a variable means both declaring a variable and assigning a value. In some programming languages these two actions can be done with one command, while in others this takes two commands.

see also declare, assign

inkjet printer *see* printer

Inkscape

Inkscape is the name of free software that you can use to make and edit vector images.

see also vector graphics

input

1 (*also* entered data) Input means data values that are put into the computer. We sometimes say that the values are 'entered'. Input is often stored as a named variable.

see also assign

2 (*also* enter) To input data means to put it into the computer, for example by typing characters on the keyboard.

see also input device

input check (*also* data check)

An input check is a double-check of data before it is entered into the computer and used in software. An input check stops user errors from crashing a program.

see also validation, verification, crash, error

input device

A input device is a device used to enter data into a computer. Manual input devices are operated by a person (such as a keyboard). Automatic input devices can read inputs more quickly than a person (such as a barcode reader).

see also input

input error *see* error

input function
(*also* input command, input keyword)

An input function is a type of program command that lets the user enter a value, for example by typing it. An input function may look like this: `variable = input()`. The value input by the user will be stored in the variable.

insert

To insert content means to add it into a document or other file. The new content does not overwrite old content.

see also overwrite

Instagram

Instagram is the name of a social media service that lets you share photos and videos. It is owned by Facebook.

install

When you install software on a device, you copy the software into storage. Then you can access and run the software on that device. An example is downloading a game from the internet onto your phone.

see also uninstall

› installer

An installer is software used for copying an app or program into storage.

instruction

In computer programming, an instruction is another word for a command. An instruction tells the computer to do something. Instruction can also mean one line of machine code.

see also command, operation, statement, code

instruction cycle *see* fetch-execute cycle

integer *see* data type

integer division *see* div

Intel

Intel is the name of a large US company that makes microprocessors, CPU and other chips.

see also microprocessor, chip, CPU

intelligent terminal *see* terminal

interactive

If a feature of a program or website is interactive, that means the user can work with it or change it. Interactive content can include quizzes, games and discussions.

interactive whiteboard

An interactive whiteboard is a board attached to a computer that can be used for input and output. Input may be through a light pen, that can write directly on the board. Interactive whiteboards are often used by teachers.

see also light pen

interface (*also* user interface)

An interface is what connects two things. When you use a computer, the interface is what lets you work with the computer. This includes the screen display and also input devices such as the keyboard and mouse.

internet

The internet is a world-wide network of computers connected by communication links. There are many services available on the internet, including email and the Web (world wide web).

see also email, world wide web

internet abbreviations

Internet messages are often short and text-based. Abbreviations are used to make messages quicker to type. Examples include: **LOL** (laughing out loud), **brb** (be right back), **smh** (shaking my head), **til** (today I learned), and **j/k** (just kidding).

internet cafe

An internet cafe is a cafe where people can use the internet. It has spare computers you can use. You can use the computers for free if you buy a coffee.

internet bot *see* bot

internet cookie *see* cookie

internet forum *see* forum

Internet of Things

Devices that are not computers can have internet connections. The network made of these devices is known as the Internet of Things. Devices such as a fridge, a central heating system or a heart monitor could have their own internet connection. People can use the connection to find things out (for example, to see if the fridge is empty) or to send instructions (for example, to turn the central heating on).

Internet Protocol address *see* IP address

internet server

An internet server is a computer that is connected to the internet and can hold a web page or other internet content (web hosting). Internet users can read content hosted on an internet server.

see also web hosting

interpreter

An interpreter is a piece of software that translates source code into instructions for the computer or machine code. An interpreter translates instructions one at a time and the computer executes each instruction immediately. The interpreter does not save any machine code and does not make an executable file. It simply goes on to the next instruction.

see also code, compiler, translate

> interpret

To interpret source code means to give instructions for the computer to turn it into machine code.

intranet

An intranet is a small network that uses web pages like the internet but is not world-wide. It is usually limited to users in a single organisation.

see also internet

WORD BUILD

> extranet

An extranet is similar to an intranet but people outside the organisation can connect to it, for example the customers of a business.

iOS

iOS is the name of an operating system made by Apple. It is used on mobile devices such as phones.

see also operating system, Apple, iMac

IP *see* TCP/IP

iPad

An iPad is a tablet made by Apple.

IP address (*also* Internet Protocol address)

The IP (Internet Protocol) address is a number-only address for a web page. It is used by the computer systems that make the internet work. An ordinary internet user does not need to know IP addresses and can use the URLs instead.

see also TCP/IP, URL

iPod

An iPod is an MP3 player made by Apple.

IPR

IPR stands for *Intellectual Property Rights*. Most countries have laws protecting IPR. This means that you cannot copy other people's work without permission. Computers make it easy to copy, so IPR is very important nowadays.

see also copyright, piracy

IRL

IRL stands for *in real life*. This is an internet abbreviation used to talk about something that happens that is not on the internet.

see also F2F

ISP

ISP stands for *Internet Service Provider*. An ISP is a company that helps you to connect to the internet and offers services like web hosting.

IT

IT stands for *Information Technology*. Computers are used to turn data (facts and figures) into information (data that has been organised and processed to make it more useful). Information Technology is the use of computers to do this.

see also ICT

iteration

If something happens many times, each time is called an iteration. Iteration means almost the same as repetition, but when we talk about computer processes we usually include the first time the thing happens.

see also loop, program structure

› iterate

To iterate is to keep repeating a process again and again.

iterative structure *see* loop

iTunes

iTunes is the name of a piece of software made by Apple. It lets you buy digital content, especially music, and lets you hear or see the content. It works through digital devices, smartphones and computers.

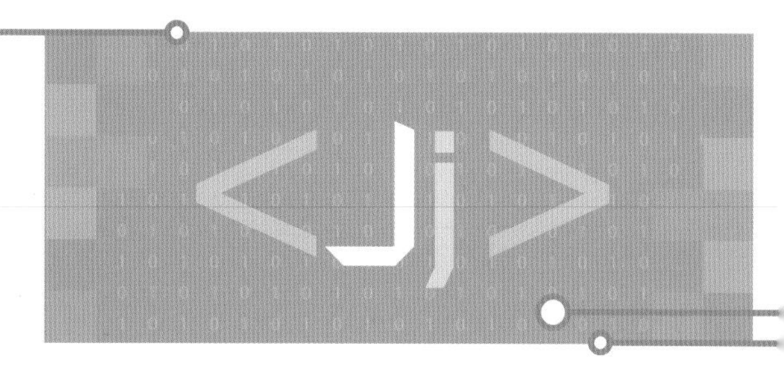

joystick

A joystick is an input device used in gaming. It is a positioning device that lets the user control movement in the game.

see also game

junk mail *see* spam

justified *see* aligned

K *see* kilo

Kbps *see* bps

key

1 A keyboard has many keys on it. When you press a key it moves down. This makes an electrical connection and sends a signal to the computer. The computer knows what key you have pressed.

2 (*also* legend) A key tells you what the colours stand for in a graph or map.

keyboard

The keyboard is a piece of equipment that you use to enter information or commands into a computer. It is one of the main ways that we interact with the computer. There is a standard keyboard layout for each country.

see also virtual keyboard, qwerty

keyboard shortcut

A keyboard shortcut is a quick way to make the computer do something using the keyboard. Most keyboard shortcuts are done by pressing the Control key and another key at the same time.

WORD BUILD

> Ctrl-Alt-Del

Ctrl-Alt-Del means press three keys at the same time, the Control key, the Alt key and the Delete key. If you press all these three keys together your computer will reboot. It is pronounced 'Control Alt Delete'.

see also keyboard, reboot, cut and paste

> undo

Undo means reverse the last action you did. The keyboard shortcut for undo is Ctrl-Z.

key field

A key field is a field in a data table that lets you tell the records apart. A code number or ID number may be used. So, for example, two students listed in a data table might have the same name but they will have different ID numbers.

see also field, database, data table, primary key, foreign key

keypad

A keypad is a small keyboard, usually with numbers only.

keyword

1 (*also* search term) A keyword is a word you enter into a search engine. The search engine will find web pages that include the keyword.

see also search engine

2 (*also* command word, reserved word) A keyword is also part of a programming language. Each keyword has a defined meaning in that language. Using a keyword makes the computer carry out an action. Keywords cannot be used as identifiers for that reason they are also called reserved words.

see also predefined function

a b c d e f g h i j k l m n o p q r s t u v w x y z

A B C D E F G H I J K L M N O P Q R S T U V W X Y Z

Kibi (*also* Ki) *see* order of magnitude

kilo (*also* K)

Kilo means a thousand. If you put kilo in front of a unit, it means a thousand of that unit. For example, a kilobyte (KB) is a thousand bytes.

see also order of magnitude

Kindle

Kindle is a type of e-reader made by Amazon.

LAN

LAN stands for *local area network*. This is a type of network that connects computers that are close together, for example in the same building.

see also WAN, network

landscape *see* page orientation

laptop

A laptop is a portable computer that has the keyboard and screen built into it.

see also portable

laser printer *see* printer

layout

Layout is the way that content, including the text and images, is organised on a page.

LCD

LCD stands for *Liquid Crystal Display*. LCD is a way to make a flat screen display and is used in many output devices. LCD does not make light, so an LCD display needs to have a light shining behind it.

see also screen

LED

LED stands for Light-Emitting Diode. An LED works like a tiny light bulb, but uses less electricity. Electricity goes into the diode. Some of the electricity passes through and some of the electricity is turned into light.

see also LCD, OLED

legend *see* key

light pen

A light pen is an input device, shaped like a pen, that you use by pointing it at or writing on a computer screen.

see also interactive whiteboard

like

1 (*also* favourite) A like is a way to mark messages in social networks. You mark other people's messages when you read them and want to show that you agree with them or like them. Likes are sometimes called favourites.

2 To like a message on a social network is to add a mark that shows that you agree with it or like it.

linear search *see* search algorithm

line break (*also* paragraph mark)

A line break makes a new line appear in a document. Word-processing software puts text into lines. Sometimes you want to make it start a new line, for example to make paragraphs in your document. You use the Enter key to make a new line.

link *see* hyperlink

linked list *see* data structure

Linux

Linux is the name of an operating system that is available for free and open-source. Linux is mainly used by experienced computer users.

see also operating system

list *see* data structure

load

To load a file or software means to copy it from storage to memory, so that it is ready to use.

see also storage, memory

local search

A local search is a search made on one computer or one web page.

see also search, query

local variable

In programming, a local variable is a variable that can only be used inside a single subroutine.

see also variable, subroutine, global variable

location

Computer memory is divided into areas called locations. Each location has an address. The computer uses locations to store data. A variable is a named location.

see also file location, memory, storage location, variable

lock (*also* lock-screen)

To lock your computer is a way to protect it if you have to leave it. The screen is hidden. The computer will not work until you enter your password again.

see also log out, log off

log file

A log file is a record of all the events that happen on a system, such as all uses of a network.

logical error *see* error

logical operator

In programming, a logical operator is an operator that joins Boolean values together to make a new value. The main logical operators are AND, OR and NOT.

see also Boolean, operator, relational operator

WORD BUILD

AND

The operator AND joins two Boolean values together. If both values are True, the result is True. Otherwise the result is False.

OR

The operator OR joins two Boolean values together. If both values are False, the result is False. Otherwise the result is true.

NOT

The operator NOT can be placed in front of any Boolean value. It reverses the value of an expression. True becomes False and False becomes True.

logical test

In programming, a logical test is an expression that has the value True or False. A logical test typically compares two values by using a relational operator. The comparison is either true or false.

see also relational operator

logic circuit

A logic circuit is an electronic circuit made of logic gates joined together.

see also gate

logical error *see* error

logic gate *see* gate

log in (*also* log on)

When you log in to a computer system you connect to your user account. You usually have to give your username and password, or you may use biometrics.

see also password, biometrics

login

1. Login means the things you do to gain access to a computer system.
2. Your login is the name you use to gain access to a computer system.

Login

username

password

Remember me

forgot your password?

Login

Logo

Logo is the name of a programming language used to teach programming to young learners. Logo commands control a floor turtle or turtle graphics.

see also high-level language, programming language, turtle, turtle graphics

log off

To log off means to break your connection to a computer system.

WATCH OUT!

You should log off if you stop work, even if it is for only a few minutes. If you leave the computer logged on, someone might use it pretending to be you. For example they might send an email from your account.

OK

log on

1 To log on to a computer system means to log in to it.

see also password, biometrics

2 People also use log on to mean use or look at a website or other computer system, without giving a username or password. For example, someone might say they 'logged on' to a website when they simply mean they started to look at the website.

see also password, biometrics

› logon

1 Logon means the things you do to gain access to a computer system.

2 Your logon is the name you use to gain access to a computer system.

loop (*also* iterative structure, iterating structure)

In programming, a loop is a program structure that repeats. The commands inside the loop may repeat many times. Each time the commands are carried out is called an iteration. A loop is stopped by an exit condition. The different types of loop have different exit conditions.

see also exit condition, iteration

condition-controlled loop (*also* conditional loop)

A condition-controlled loop is controlled by a logical test. The number of iterations is not set in advance.

counter-controlled loop (*also* fixed loop, finite loop)

A counter-controlled loop iterates a set number of times. A counter variable counts the number of iterations. When the counter reaches a set value the loop stops.

see also counter variable

forever loop (*also* infinite loop)

A forever loop is a loop that has no exit condition. There is no way to stop the loop and it will keep running until the program stops.

lossless compression
see compress

lossy compression *see* compress

lower case *see* case

low-level language

A low-level language is a type of programming language. The commands in a low-level language match the instructions that control the processor. This type of program is harder to write and understand than one written in a high-level language. Assembly language and machine code are types of low-level language.

see also assembly language, high-level language, code

low-resolution (*also* low-res, lo-res)

A low-resolution image is made of a few large pixels. The image is less detailed, but the file size is small.

see also resolution, high-resolution

LOW-RES

HIGH-RES

a b c d e f g h i j k l m n o p q r s t u v w x y z

M *see* mega

machine code *see* code

machine learning

Machine learning is a way to program a computer. Instead of just giving it instructions, you also give it examples, and the computer can learn what the examples have in common. For example, it could study a set of photos and learn to recognise photos of human faces.

see also neural network

machine-readable

If something is machine-readable, it can be input automatically into the computer. A person does not have to type the data.

see also input

macOS

macOS is the name of an operating system made by Apple. It is used on computers such as laptops.

see also operating system, Apple, iOS, iMac

macro

A macro is a software shortcut that stores several commands. When you run the macro all the stored commands are carried out.

magnetic storage (*also* magnetic media)

Magnetic storage is a way to store files. The file is held electronically in memory. The electricity can be turned into magnetism, which is held on a disk or on tape.

mailing list

A mailing list is a group of email contacts. If a message is sent to the mailing list, everyone on the list will get the email.

mail merge

Mail merge is a word-processing feature that lets you create copies of a letter to send to a large number of people. Mail merge takes the text of a letter and merges (combines) it with a list of names and addresses. The result is a number of letters, one addressed to each name on the list.

mainframe (*also* mainframe computer)

A mainframe computer is a very large computer. It is usually placed in a room on its own. Many people can use the computer at the same time. They work at terminals or workstations connected to the mainframe computer.

see also terminal, workstation, multi-user system

main memory *see* memory unit

malware

Malware is short for *MALicious softWARE*. This is software made to cause harm. It hides so that it cannot be easily deleted from the computer. The most common sort is a virus. Antivirus software is designed to help find and remove malware from a computer.

see also virus

WORD BUILD

> trojan (*also* trojan horse)

A trojan is a type of malware that hides inside other software. If you download a free game, for example, it might have a trojan virus.

> worm

A worm is a type of malware. It is a hidden file on the computer.

Mars Rover

A Mars Rover is a robot that is used on Mars. It is not remote-controlled from Earth but is designed to work out its route on its own. There are many Mars Rovers, but some are inactive now.

maximise *see* window

Mbps *see* bps

Mebi *see* order of magnitude

mega (*also* M)

Mega means a million. If you put mega in front of a unit, it means a million of that unit. It is abbreviated to M. For example, a megabyte (MB) is a million bytes.

see also order of magnitude

memory

The memory of a computer is its electronic storage. The memory holds the data and instructions that the computer is using at any particular moment.

see also RAM, ROM

memory address

Every part of computer memory has its own address, which the processor can use to get the data it wants. The addresses are binary numbers.

memory card

A memory card is a card with a small amount of flash memory. It is used for file storage.

see also flash memory, storage

memory stick *see* pen drive

memory unit (*also* IAS, main memory)

The memory unit is the part of the processor where the Random Access Memory (RAM) holds the data and instructions that the computer uses.

see also RAM

A B C D E F G H I J K L M N O P Q R S T U V W X Y Z

menu

A menu is a list of choices. You click on the choice you want.

see also software window, online form

WORD BUILD

> context-sensitive menu

If a menu is context-sensitive, then the choices on the menu will change to fit the task you are doing.

> menu bar

Some apps have a menu bar at the top that lists the names of menus. You click on the name of the menu you want to see.

> types of menu

Some menus are shown in full, but many are hidden until you open them. For example, they might be 'drop-down' or 'pop-up'. You click on the menu name to see all the choices.

File Edit View Insert Tools

New
Open
Close
Save
Save As . . .

Exit

metadata

Metadata is data about data. It is data that describes content. For example, the metadata for a website tells you what the site is about.

micro-

Micro- means very small. You can put micro- in front of another word. For example, a microprocessor is a very small processor.

microchip *see* chip

microcomputer

A microcomputer is a computer built around a single microchip processor. It is made for one person to use.

see also personal computer

microprocessor

A microprocessor is a computer processor made from a single piece of silicon. Modern personal computers are built around a microprocessor.

see also process, chip

Microsoft (*also* MS)

Microsoft is the name of a large technology company. Microsoft make MS Office, Windows and the Edge browser.

see also Windows, MS Office, Edge

microwave link

A microwave link is a way of sending digital signals by using microwaves. Microwaves are energy waves, similar to radio waves.

⚠ WATCH OUT!

The microwave signals used to link computers are not powerful like the ones used in a microwave oven. They are not dangerous.

OK

MIDI

MIDI stands for *Musical Instrument Digital Interface*. A MIDI file stores the instructions and notes that make a piece of music. A MIDI file can be read by an electronic instrument. It can then play the music.

milestone

A milestone is a date in the future used when planning a project. You plan to get a particular task done by that date. When the milestone date comes round, you can check that your project is still on track.

minimise *see* window

MIPS

MIPS is short for *millions of instructions per second*. Like flops or Hertz it is a measure of computer speed. The larger the number of MIPS, the faster the computer.

see also flops

mirror site

A mirror site is an exact copy of a website. You might look at the mirror site if the main site is too busy.

MMORPG

MMORPG stands for *massively multiplayer online role-playing game*. This is a multiplayer game with thousands or even millions of players. The players play in a huge shared world. World of Warcraft is a large and successful MMORPG.

see also RPG

mobile communication

Mobile communication is a way to link devices using a wireless connection. The device does not have to be plugged in and you can move around holding the device.

see also device, wireless

mobile phone (*also* cell phone)

A mobile phone is a phone with a wireless connection. It can send and receive messages without being plugged in.

MOD (*also* modulo)

In programming, MOD is an arithmetic operator that gives you the remainder of a division sum. In mathematics it is called modulo.

see also arithmetic operator, div

model

A model is a simplified version of a complex real-life system. A program could not match all the details of a real-life system, but a simplified model might be good enough. A program based on a model could help you to find a solution to a real-life problem.

see also abstraction

modem

A modem is a device that converts phone signals to computer signals. A modem is one of the devices that can be used to join a computer to the internet.

see also network

module (*also* modular component)

In programming, a module is a single subroutine in a program. Module is also the name for a file that stores subroutines, so they can be used in other programs.

see also code library, subroutine

modulo *see* MOD

A B C D E F G H I J K L M N O P Q R S T U V W X Y

monitor

A monitor is an output device that shows the output on a screen.

see also device, screen, VDU

monospace *see* font

Moore's Law

Moore's Law is the theory in computer science that computers will keep getting smaller and more powerful. Nobody knows if this will go on forever. Moore's Law is named after Gordon Moore, who first suggested it. Gordon Moore was one of the founders of the US company Intel.

motherboard

The motherboard is the main circuit board of a computer. It holds the main processor.

see also circuit board

motion sensor

A motion sensor is an input device that can detect movement. For example, it might be used in a burglar alarm system.

mouse

A mouse is an input device. You move it about on a flat surface. It sends a signal to the computer every time it moves. The computer moves the pointer on the screen. When you click a button on the mouse this sends another signal.

mouse-click *see* click

mouse mat

A mouse mat is a flat piece of plastic that is designed to help a mouse work properly. Sometimes a mouse does not work properly on an ordinary table surface.

see also mouse

mouseover

A mouseover is the action of moving the mouse pointer so that it touches an object, but you do not click. You may see a message, for example a tip about how to use the feature. This is used in web page design.

see also pointer

mouse pointer *see* pointer

move

To move an object is to make it disappear from one place and appear in a different place. Cut and paste is one way to move an object. Dragging is another way.

see also cut and paste

Mozilla Firefox *see* Firefox

MP3

MP3 is a file format used for storing sounds. The sounds are stored with medium quality.

MP3 player

An MP3 player is a device that plays MP3 sound files.

MP4

MP4 is a file format used for storing sound and video.

MPEG

MPEG stands for *Moving Picture Experts Group*. The group tries to standardise how audio and video files are made. This makes it easier to share them over the internet.

MS *see* Microsoft

MSN

MSN is short for *Microsoft Network*. It is a web portal operated by Microsoft.

see also web portal

MS Office

MS Office is a group of software applications made by Microsoft. You usually buy them all together.

see also Microsoft

WORD BUILD

Excel

Excel is a spreadsheet application. It is part of MS Office.

PowerPoint

PowerPoint is a presentation application. It is part of MS Office.

Word

Word is a word-processing application. It is part of MS Office.

see also spreadsheet, presentation, word processor

multimedia

Multimedia means all types of digital content, including images, sounds and video.

multiplayer gaming

Multiplayer gaming is a type of gaming that allows you to play a computer game with other people. You mightcompete with them or co-operate to solve a problem. Those who play together online share access using communications links.

multitasking

Multitasking is when one computer does many different tasks at the same time. The computer is in fact swapping very quickly between the different tasks, but it does this so quickly you do not notice any delay.

multi-user system

A multi-user system is one in which many users all share the same computer, for example a mainframe computer. The computer swaps very quickly between the different users.

see also mainframe

name *see* identifier

nano-

Nano- means one billionth. For example, a nano-metre is one billionth of a metre. That is microscopic in size. Putting nano- in front of any word means 'very small'.

nano-bot

A nano-bot is an idea for a very small robot, perhaps as small as a piece of dust. A self-replicating nano-bot could make more nano-bots, though this has not yet been invented.

see also robot

natural language interface

Natural language means ordinary speech. A perfect natural language interface would allow you to communicate with a computer exactly as if you were talking to a real person. This has not been invented yet.

see also speech recognition

navigate

To navigate means to find your way. Navigating the web means finding the web page you want to look at.

> navigation

Navigation means finding your way around online or on a website.

nested

In programming, structures can be nested. This means that one structure is put inside another structure. Nesting is often shown using double indentation.

see also indent

nested if (*also* nested conditional)

In programming, a nested if is an if structure inside another if structure.

see also if statement

nested loop

In programming, a nested loop is a loop inside another loop.

see also loop

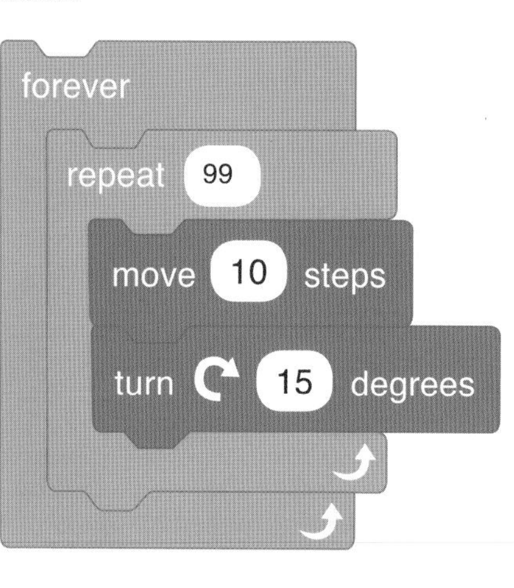

net

The internet is sometimes called the net for short. Net can also be used to mean the world wide web.

see also internet, world wide web

netiquette

Netiquette means being polite on the internet. The word is made from the word 'net' and the word 'etiquette' (rules about polite behaviour) joined together.

A B C D E F G H I J K L M N O P Q R S T U V W X Y Z

network

A network is a group of computers that are joined together by communication links. They can share data and sometimes they share software too.

WORD BUILD

> hub

A hub is a device that joins parts of a network together. It sends and receives signals from the computers in the network. Many hubs use wireless signals.

> node

Node is a general term for a place on a network. It can be the end of a connection or it can join two connections together.

> router

A router is a device that joins two networks together. Many people have a router in their home to join their home network to the internet.

> switch

A switch is a device that joins parts of a network together. Like a hub, it sends and receives signals but a switch has more processing power. A switch only sends a signal to the computer that needs to receive it. This stops the network from getting overloaded.

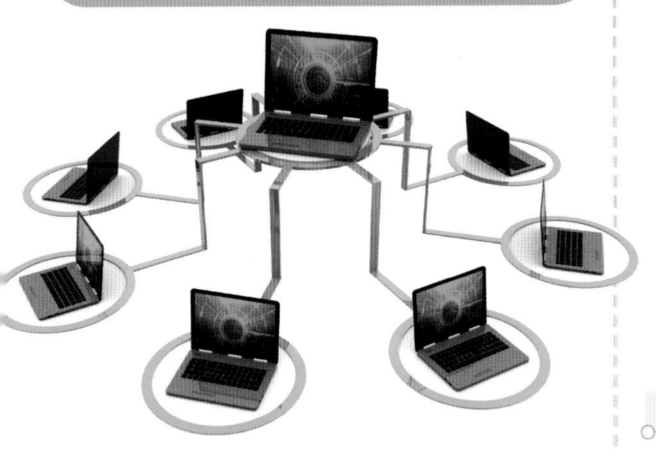

network card

A network card is an expansion card that lets a computer connect to a network. It goes inside the computer case. If you do not have one you can use a dongle instead.

see also dongle, expansion card

network server *see* server

network share *see* share

neural network

An artificial neural network is a way of using a computer processor. It is an alternative to writing programs with instructions. The connections of a neural network are designed to behave a bit like the nerve cells in a brain. It is used for machine learning.

see also machine learning

Nintendo

Nintendo is the name of a Japanese company that makes computer games, including Pokémon.

node *see* network

none *see* null

non-volatile

Non-volatile describes computer storage that does not use electricity. Data in non-volatile storage is not lost when the electricity is turned off.

see also volatile

NOT gate

In a processor, a NOT gate is a type of electronic gate. A signal goes into the gate and it reverses the signal. If the signal going in is ON the signal coming out is OFF.

see also gate, process

NOT operator *see* logical operator

NPC

NPC stands for *Non-Player-Character*. This is a game character that is not controlled by any player, but is controlled by the computer. Enemies and helpful bystanders may be NPCs.

see also character

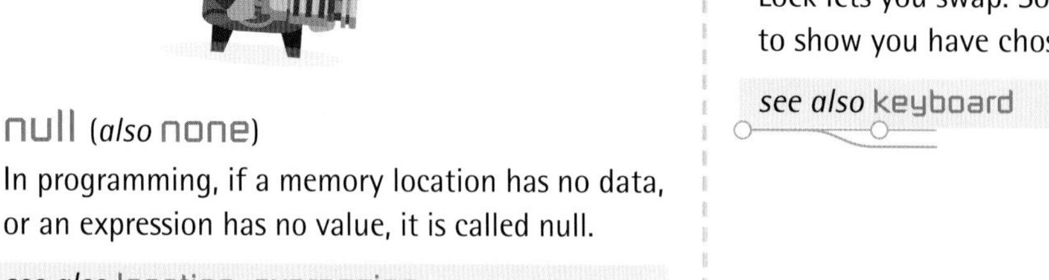

null (*also* none)

In programming, if a memory location has no data, or an expression has no value, it is called null.

see also location, expression

number base

A number base is a way of showing number values. Number values are shown using symbols called digits. There are many different number bases. Each base shows the same number value in different ways. For example, in base ten the value one hundred is shown using the digits 100. In base sixteen the same value is shown using the digits 64.

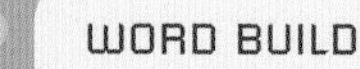

> decimal (*also* base ten)

Our normal number system is base ten. Base ten numbers are also called decimal numbers. Another term is denary numbers. Base ten uses the ten digits 0123456789.

> binary (*also* base two)

The number system used inside the computer is base two. Base two numbers are also called binary numbers. Base two uses the two digits 1 and 0.

> hexadecimal (*also* hex, base sixteen)

Computer scientists often use base sixteen. Base sixteen numbers are also called hexadecimal numbers. Base sixteen uses the sixteen digits 0123456789ABCDEF.

see also binary, hexadecimal

Num Lock key

Num Lock stands for *number lock*. On many keyboards the number keys also have arrows on them. You can swap between using the keys as numbers or using them as arrows. Pressing Num Lock lets you swap. Sometimes a light comes on to show you have chosen numbers.

see also keyboard

object

When you use the computer an object is any feature that you can work with, such as a piece of text or an image. You can select an object by clicking with the mouse.

see also select, graphics

OCR

OCR stands for *Optical Character Recognition.* An OCR reader can scan a paper document. The computer can recognise the different letters and turns the letters into a text file.

Office 365

Office 365 is a service offered by Microsoft that lets you use MS Office software in your web browser.

see also MS Office

offline (*also* off-line)

Offline means not connected to the internet. Offline activities are done without using an internet link.

see also online

OK button

A dialogue box often has an OK button. You click this button if you agree with the message or question shown in the dialogue box.

see also dialogue box

OLED

OLED stands for *Organic Light-Emitting Diode.* OLED is a technology that makes display screens. It uses a grid of LED lights to make the display. OLED is an alternative to LCD screens.

see also LCD, LED

online (*also* on-line)

Online means connected to the internet. Online activities are done via an internet link.

see also offline

online chat *see* chat

online form (*also* computer form)

A form is a way to collect information that is structured into questions and sections. An online form is a form on the computer. Online forms often have text boxes, check boxes or radio buttons, as well as a submit button.

see also text box, checkbox, radio button, submit button

online identity

Your online identity is the name you use on the internet. It might be the same as your real-life name or it might be different.

online transaction

A transaction is an event such as buying or selling. An online transaction is a transaction done over the internet.

see also transaction

A B C D E F G H I J K L M N O P Q R S T U V W X Y Z

OOP

OOP stands for *object-oriented programming*. Object-oriented programming is a way of writing programs in which the programmer defines new data structures. Each data structure has its own values, called attributes. Each data structure also has its own functions and procedures, called methods.

see also attribute, data structure, function, procedure, programming language

open

To open a file or an app means you start it up ready to work with it. The contents are moved from storage to the active electronic memory. A file or app in storage is shown as an icon. Double-clicking on the icon opens the file. It usually opens as a new window on the screen.

open source software

Open source software is a way of sharing software. The users can look at the source code as well as the machine code and can usually make changes too.

see also code

operating system

An operating system is a collection of system software. It has all the software the computer needs to start up and run the computer hardware. When you buy a computer it usually has an operating system installed already.

see also Android, iOS, Linux, macOS, Windows

operation

In programming, an operation is a single action that the computer carries out, such as doing a calculation. A command in a program tells the computer to carry out an operation.

see also command, instruction, statement

operator

In programming, an operator is a symbol that tells the computer what operation to carry out. For example, the arithmetic operator + tells the computer to do an addition.

see also arithmetic operator, assign, logical operator, relational operator

optical fibre *see* fibre-optic

optical storage

Optical storage is a way to store digital data. The data is stored as microscopic pits burned into the surface of a CD, DVD or other similar disk by a laser. A less powerful laser is used to read the pits.

WORD BUILD

> Blu-ray

Blu-ray is a type of optical storage that uses very small pits so that it can store high definition video. The pits are made with a very thin blue laser. It stores 50 GB of data.

> DVD

DVD stands for *digital video disk* or *digital versatile disk*. DVD can store more information than CD, but less than Blu-ray. It stores 4-7 GB of data.

> CD

CD stands for *compact disc*. This is a plastic disk used for optical storage, for example music storage. It stores 700-800 MB of data.

> CD-RW

CD-RW stands for *CD-rewritable*, a type of CD on which you can change the data. The plastic coating on the CD will melt so the pits disappear and new data can be added.

optimise

To optimise something means to make sure that it is working at its best.

option key *see* Alt key

order *see* sort, sort order

order of magnitude

Modern computer files are often very large, containing millions or even billions of bytes. We use 'orders of magnitude' to help us talk about big numbers in a simple way.

see also bit, byte, file size

Computer storage, processing power and transmission speed are measured using large number units called 'orders of magnitude'.

TERM	MEANING	ABBREVIATION	EXAMPLE	AS A NUMBER	AS A POWER OF 10
Kilo	Thousand	K	1000 bytes = 1 KB	1,000	10^3
Mega	Million	M	1000 KB = 1MB	1,000,000	10^6
Giga	Thousand million	G	1000 MB = 1GB	1,000,000,000	10^9
Tera	Million million	T	1000 GB = 1TB	1,000,000,000,000	10^{12}
Peta	Thousand million million	P	1000 TB = 1PB	1,000,000,000,000,000	10^{16}

NOTE:

You will see that the orders of magnitude are round numbers. They increase by multiples of 1000.

Some computer scientist use binary 'orders of magnitude' instead. These are round numbers in binary. They increase by multiples of 1024. These terms should correctly have 'bi' included in the name (short for 'binary').

SO:

1 Kibibyte = 1KiB = 1024 bytes

1 Mebibyte = 1024 kibibytes

And so on

Reality check

Although we measure storage in MBs and GBs, in reality a computer is rarely able to use every bit of its storage capacity to the full. So the size should be treated as a guideline rather than as a precise figure.

OR gate

In a processor, an OR gate is a type of electronic gate. Two electrical signals go into the gate. One signal comes out. If either or both the inputs are ON the output is ON.

see also gate, process

OR operator *see* logical operator

output

Output means taking data from the computer and putting it in a useful form. The output might be displayed for people to see, it might be sound, or it might control a process.

output device

An output device is any device used for output, such as the screen, speakers or control devices.

see also screen, speaker, control device

overclocking

Overclocking means increasing the speed of the clock timer inside the computer. This can make the computer go faster. However if you increase the speed too much it can cause problems. One example is the computer could overheat.

see also clock speed

overwrite

To overwrite content means to add it into a document or other file so that it replaces the content that is there already.

see also insert

P *see* order of magnitude

P2P

P2P stands for *peer-to-peer*. Peers are things of equal importance. A P2P network joins computers that are equally important. P2P is the alternative to a client-server network.

see also client-server

packet-switching

Packet-switching is the way data goes through the internet. The data is split into pieces called packets. Each packet makes it own way to the destination. They can go by different routes. At the end they are put back together again.

see also TCP/IP

page break

A page break starts a new page in a document. Word-processing software puts text into pages. Ctrl-Enter is the keyboard shortcut to make the computer start a new page.

see also keyboard shortcut

page formatting

Page formatting means how text and images are set out on a page.

see also format

page hit

A page hit is when a file on a web page has been read over the internet. The number of hits can be much bigger than the number of visits to the page. For example, each picture on the page is a different file. This means that one page view can give many page hits. About half of all page hits are from bots or web crawlers.

see also page view, bot, web crawler

page orientation

Most paper is rectangular, with one side shorter than another. Page orientation is how the page is turned when the document is printed.

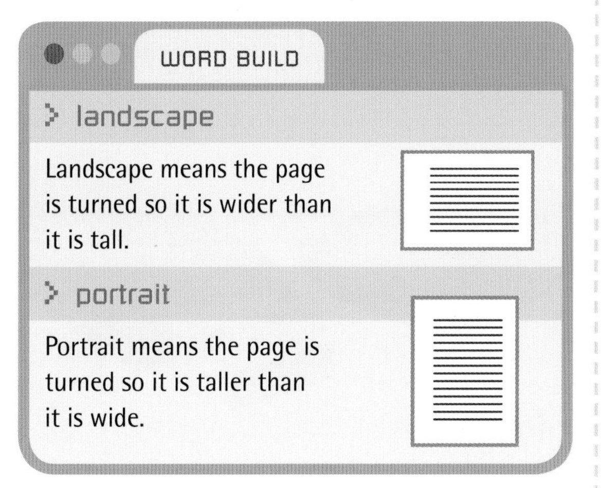

page view

A page view is when someone has looked at a web page over the internet. A person who looks at a web page is called a visitor. Whether visitors include bots or only humans depends on how the visit is recorded. Different sites use different methods to record pageviews.

palm print *see* biometrics

paper jam

A paper jam is when a piece of paper has got stuck inside a printer. You usually have to open the cover and pull the paper out.

see also printer

paragraph mark *see* line break

parallel

Things that happen in parallel happen at the same time rather than one after the other. Parallel processing is when several instructions are carried out at the same time.

see also serial

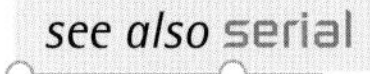

parallel port

A parallel port is one that allows parallel transmission so that the bits of a signal are sent at the same time.

see also port

parallel processing

Parallel processing is when a computer shares a task between many different processors inside it. The different processors can share parts of a task between them. That means the task gets finished more quickly.

parallel transmission

Computer data is stored in bytes. Each byte has eight bits. In parallel transmission all eight bits are sent at the same time, typically along eight different wires. This is a fast way to transmit data between devices or parts of a device.

see also serial transmission

parameter (*also* argument)

In programming, a parameter is a value passed to a subroutine. The name of a subroutine is followed by round brackets. A parameter can go inside the brackets. Here is an example: `find_root(25)`. This subroutine is called `find_root` and the parameter is the value `25`. The parameter value can be used by the commands inside the subroutine.

see also subroutine, procedure, function

parentheses *see* brackets

A B C D E F G H I J K L M N O P Q R S T U V W X Y Z

password

A password lets you log in to a computer system. It is a secret input that you use to authenticate who you are.

see also login

commonly used password

The most commonly used passwords are: the word 'password', the string 'qwerty', the numbers '1234'. Passwords like these are not good ones to use because they are easy to guess.

strong password

A strong password is one that is hard to guess. It is not a well-known word and includes characters that are not letters, such as numbers and symbols.

weak password

A weak password is one that can be easy to guess. Many weak passwords are short, made of letters only and are often a well-known word or name. People sometimes use their birthday. That is easy to guess so it is a weak password.

WATCH OUT!

Do not share your password with anyone. Do not write it down.

OK

patch

A software patch is a kind of add-on that is used to fix a problem.

see also add-on

pause

If software pauses, it stops running and waits for a signal from the user before it continues.

PC *see* personal computer

PDF

PDF stands for *Portable Document Format*. It is a file format developed by Adobe to display text and images in a fixed layout that is independent of any particular hardware, software or operating system. A PDF file can be read by any computer and has the file extension .pdf.

see also Flash

pen drive (*also* flash drive, memory stick)

A pen drive is a piece of flash memory attached to a USB connector. It is used for storage.

see also storage

Pentium

Pentium is a brand of microprocessor made by Intel.

see also microprocessor

performance

The performance of a computer is how much work it can do. A high performance computer can do a lot of work in less time than a low performance one.

peripheral

A peripheral is any device plugged into a computer. This could either be an input device or an output device, for example a mouse or a monitor.

see also input device, output device

permission
Permission is the right to access data. You may be given read-write access or read-only access.

see also access

personal computer (*also* PC)
A personal computer is a computer made for an individual to use. It is sometimes called a PC.

see also microcomputer

personal data
Personal data is data that relates to a person and can identify the person, such as their name, address or facial features. Personal data is protected by law.

see also data protection

peta (*also* P) *see* order of magnitude

petaflop
A petaflop is a unit of computing speed. It means that a computer can do a thousand million million operations in one second, which is extremely fast.

see also flops

pharming
Pharming is a way to trick people on the internet. It means making a trick website, such as a site that looks like the web site of a real bank. Some people will then enter their bank password. Now the thieves know the person's password and can steal their money.

see also identity theft, hacking

phishing
Phishing is a way to trick people on the internet. It means sending a trick email, such as an email that looks like it comes from a real bank. Some people will email back with their personal details.

see also identity theft, hacking

Photoshop
Photoshop is a piece of software that is used to edit photos. It uses bitmap graphics.

see also bitmap graphics

ping
A ping is a signal you can send on a comms link. It goes out and comes back. It is used to test the link.

see also comms

piracy
Piracy means making illegal copies of music, TV shows, games and so on.

see also IPR, copyright

pixel
A pixel is a single point in an image. In a screen image it is a point of light. In a printed image it is a spot of ink.

see also bitmap graphics, freeware

pixelated
A pixelated image is one in which you can see the dots that make up the image. This means that the image cannot be seen clearly.

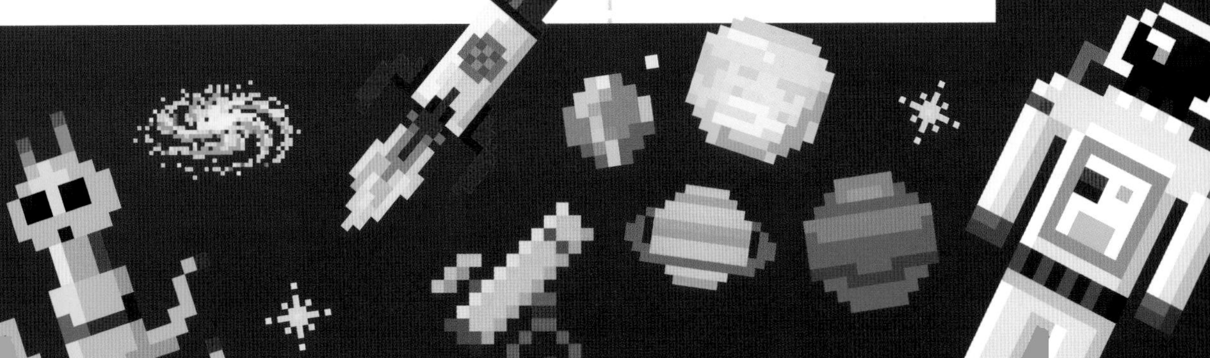

A B C D E F J K L M N O P Q R S T U V W X Y Z

platform

A platform means all the computer parts needed to run some software. This might include the hardware and operating system, together with an internet connection.

see also hardware, software

platform game

A platform game is a game genre. It is 2D (flat-looking). The character runs along and jumps onto platforms. Some popular platform games are Mario and Sonic.

see also game genre

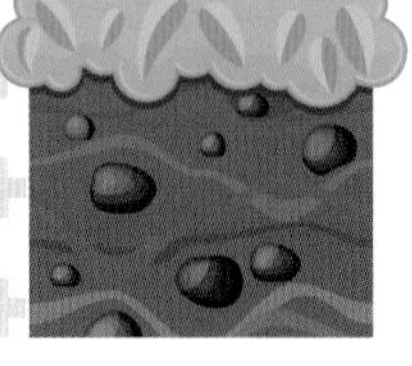

plotter

A plotter is an alternative to a printer. It is a device that produces images on paper. The plotter makes the image by drawing with a pen. The computer controls the arm that moves the pen.

plug and play

Plug and play means you can plug a device into your computer and use it right away. The computer knows how to connect to the device. Common standards of design mean most devices nowadays are plug and play.

see also standard

plug-in *see* add-on

podcast

A podcast is a show you can listen to over the internet. A podcast is a series of episodes. You can download the episodes and listen to them. You can also get video podcasts.

pointer

The pointer (or mouse pointer) is an arrow on the screen. As you move the mouse the pointer moves too. You use it to select and click.

see also mouse, pointer device

pointer device

A pointer device is an input device. It lets you select and move objects on the screen. A mouse is a pointer device.

see also mouse, touchpad, touchscreen

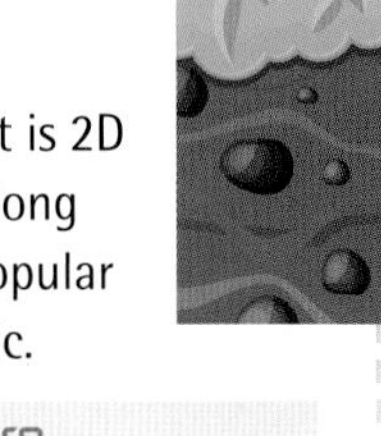

port

A port is a socket on a computer that you can use to plug in a device. There are many kinds of port. You must have the right kind of port on your computer or you cannot use the device.

see also USB, HDMI, DVI, Ethernet, serial port, parallel port

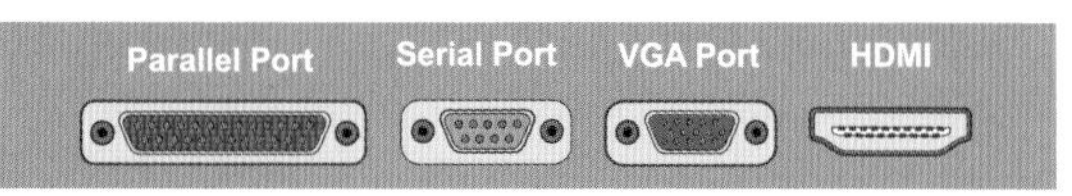

portable

1. If something is portable, you can carry it. Some computers are portable, while some are not.
2. Portable software can run on more than one platform. It is easy to start using portable software from any system or device.

see also platform

portable hard drive

A portable hard drive uses magnetic or flash storage. It is a plug-in device that you can move between computers.

see also storage

portrait *see* page orientation

POS *see* EPOS

post

1. A post is an entry made by a user on a blog or social media site.
2. If you post an entry on a blog or social media site, you upload it there.

see also social media

PowerPoint *see* MS Office

predefined function

A predefined function is a function that you do not have to write. It comes as part of a programming language. You can use the function in your code.

see also function

presentation (*also* slideshow)

A presentation is a series of slides that you use as you give a talk. There are different apps you can use for creating presentations, such as PowerPoint.

see also slide

primary key

A primary key is the main key field for a data table. It is called a 'unique identifier'. Only one record has that key value.

see also key field, relational database, data table

printed circuit *see* circuit board

printer

A printer is an output device that produces a copy of a document on paper.

inkjet printer

An inkjet printer is a medium quality printer. It works by spraying a jet of ink onto paper. It can be black and white or colour.

laser printer

A laser printer is typically a larger and faster printer than an inkjet printer. This is the type of printer used in the workplace. A fine dust called toner sticks to a cylinder by static electricity. The paper rolls past and picks up the toner.

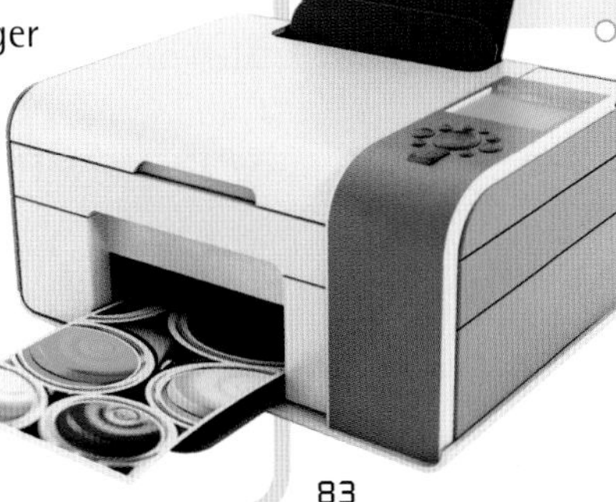

printing

Printing is a form of output from the computer. Content such as documents and images are output onto paper. This is a good way to keep a permanent record of the content.

printout *see* hard copy

print preview

Print preview is an on-screen display. It shows you what a document will look like when it is printed. You can check this before you start printing to paper.

print screen

Print screen is a key on the keyboard. It is usually marked 'PrtScn'. Pressing this key takes a screenshot of the screen.

see also keyboard, screenshot

priority

Priority means how important something is. When a computer is multi-tasking, tasks with high priority are given more time so that they are completed more quickly.

privacy

Privacy is the fact that data is protected. Only people who have permission can access the data. People without permission cannot see it.

see also permission, security, access

procedure

In programming, a procedure is a type of subroutine. Unlike a function, a procedure does not create a new value.

see also function, subroutine

a b c d e f g h i j k l m n o p q r s t u v w x y

A B C D E F G H I J K L M N O P Q R S T U V W X Y Z

process

1 To process means doing work to change or transform something. A computer is a machine that processes data to make new information. The computer does this by organising and changing the data. It uses the rules of maths and logic.

2 A process is a set of actions to make a change. A computer program holds instructions to carry out a process. Each time you run the program that starts a new process.

processing

When the computer is working we say it is processing. Data processing or information processing is a way of describing the work of a computer system.

processor

A processor is an electronic device that takes electronic signals and changes them in a controlled and logical way. A processor is at the heart of a computer system.

see also CPU, microprocessor

processing speed *see* clock speed

program

A program is a collection of instructions. A programmer writes the program as source code. It is turned into machine code. Then the program is ready to run. When you run the program the computer follows the instructions.

see also code, run, software

programmable

Something is programmable when you can change the program that controls it. If a device is programmable, you can change what it does.

programming language (*also* computer language)

A programming language is the language used to write source code. There are many different programming languages, such as Python, Scratch, Java and C++.

see also code, high-level language, low-level language

program structure

There are three main structures used in programs: sequence, selection (if structure) and iteration (loop structure). In a visual programming language a program structure is shown as a block.

WORD BUILD

iteration

Iteration means repeating something. A loop is an iterative structure. The commands inside a loop are repeated. Each repeat is called an iteration.

see also loop

selection

Selection is when the computer chooses between actions based on a test. In most programming languages, this is done by an if structure.

see also if statement, logical test

sequence

A sequence is when the computer carries out program commands one after the other. The computer follows the sequence of commands in the program.

prompt

A prompt is a message added to an input command. It tells the user what data they have to enter.

protocol (*also* communication standard)

A protocol is a standard way of communicating, such as the pattern of bits that starts and ends a signal. Shared protocols help computers to communicate.

see also standard

pseudocode

Pseudocode is a way of setting out an algorithm in plain language. It sets out the steps to solve a problem using words that look very like source code. But a pseudocode algorithm is not designed to be read by a computer and does not work as a real program does.

```
if age < 60 then
   ticket = 12.50
else
   ticket = 29.99
endif
```

public domain

If content is in the public domain, it can be used and copied. It is not covered by IPR and nobody owns it. For example, work done by people who have been dead for more than 70 years is in the public domain. The Gutenberg Project is a website that has books that are in the public domain.

see also IPR, copyright

publish

If you publish content on the internet, you put the content on a web server so that people can look at it using an internet connection.

Python

Python is a text-based programming language with a large community of programmers. There are many modules you can add to a Python program. This means that you can use functions that other programmers have made.

see also high-level language, programming language

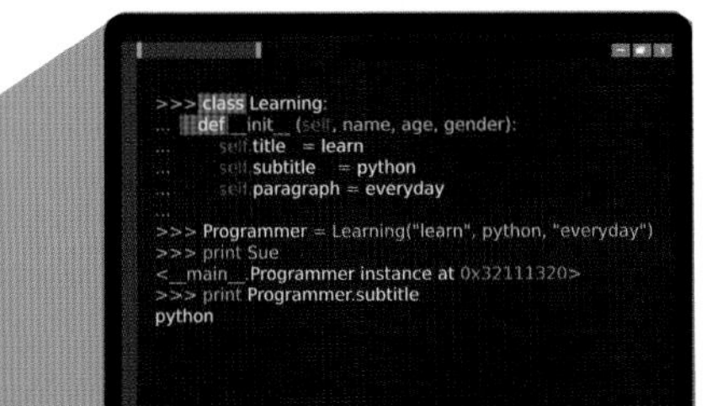

quantum computer

Quantum physics is about particles smaller than an atom. A quantum computer is one that uses these very small particles. Quantum computers might become common one day, but at the moment the idea is early in its development.

query

A query is a way to find data in a database. The query describes the data you are looking for. Then the software finds records that match the query.

see also database, SQL

queue

In computing, a queue means any list of jobs or tasks that are waiting to be dealt with. When an item is added it goes to the back of the queue. The items at the front have priority. For example, items at the front of a print queue will be printed first.

see also priority

qwerty

Most computers in English-speaking countries use a standard keyboard layout. The first letters on the top line are Q W E R T Y. This is why it is sometimes called a qwerty keyboard.

see also keyboard

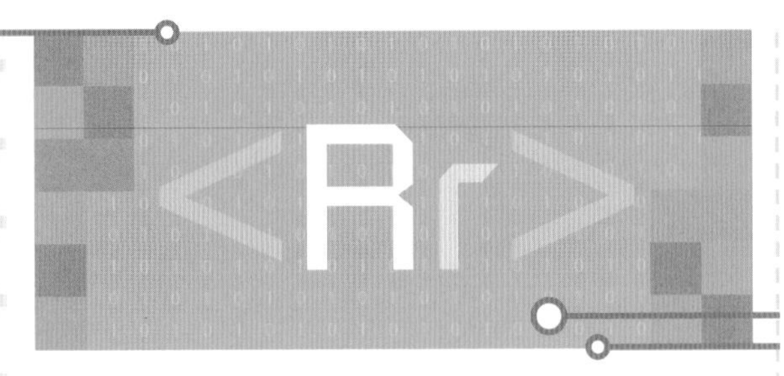

radio button

Radio buttons are used on an online form. You get a list of options with a button or a circle next to it. Each button stands for one choice and you click on the one you wish to select. There are always two or more choices and only one button in the group can be selected.

see also checkbox, online form

What is your favourite animal?

- ○ Lion
- ◉ Chameleon
- ○ Monkey
- ○ Frog
- ○ Meerkat
- ○ Tiger

RAM

RAM stands for *Random Access Memory*. It is the name for the electronic memory inside a processor. RAM is volatile.

see also memory unit, volatile

random access

Random access describes memory or storage. Random means that the computer can access any part of the storage, i.e. any random location.

see also sequential access, RAM

range

A range is part of a sequence of things, such as numbers or records. You can identify a range by saying the first and last values of the range.

raster graphics *see* bitmap graphics

read

1 When a computer reads data, the data is loaded into the computer's memory (RAM) and can be processed or used by the computer.

2 Data can be read from an input device or from storage.

3 The computer can also read a file. This means that the whole file is loaded into RAM and can be used.

see also load, RAM, storage

readable *see* code quality

README file

A README file is a text file that has help or advice about other files, such as software programs. You open the file to see what it says.

see also help, document

real, real number *see* data type

real-time

A real-time operating system is one that responds at real-life speed with no delay. This would be needed to control a self-drive car, for example, which must respond immediately to dangers.

see also operating system

reboot

To reboot a computer means to shut it down and start it up again. Any work that is not saved to storage will be lost. If a computer is stuck or frozen, you might need to reboot it.

see also boot, keyboard shortcut, restart

recent files

Recent files are the files that you have just opened or saved. They might be in different storage locations. A file manager can show recent files.

see also file manager

recondition

To recondition a machine means to repair it so that it is almost as good as new. Buying a reconditioned computer is cheaper than buying a new one.

record

1 A record in a data table stores all the information about one person or thing. One row of a data table stores one record.

see also data table

2 To record something is to store data about it. For example, you can record a video, a sound or some facts.

recursion

In programming, recursion is when the name of a subroutine is included as one of the commands inside that same subroutine. The subroutine calls itself. This makes the subroutine repeat over and over again.

see also subroutine, call

recycle bin

The recycle bin is a folder where files go when you delete them. You can see the icon on the desktop. Looking in the recycle bin gives you a chance to restore files that you have deleted by mistake.

see also restore

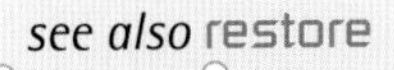

Reddit

Reddit is the name of a website for online discussions. On the Reddit site there are lots of different forums called subreddits. Each one is about a different topic.

see also forum

refresh *see* reload

relational database

A relational database is made up of data tables that are linked together by key fields. Because of the links data can be combined from more than one table.

see also database, data table, key field

relational operator

In programming, a relational operator is used to compare two values and create a logical test. Some of the most common relational operators are:
= (equal to; sometimes written as ==),
!= (not equal to; sometimes written as ≠ or <>),
> (greater than), < (less than).

see also operator, logical operator

relative cell reference *see* cell reference

a b c d e f g h i j k l m n o p q r s t u v w x y z

reload (*also* refresh)

In computing, to reload or to refresh something means to load it again. For example, you can reload a web page. The computer will go back to the web server and will get the data again.

see also load, update

remote access

Remote access means a storage location that is away from the computer. The computer can access the storage through a comms link.

see also comms, cloud-based

remote content

Remote content is content stored far from your computer. You can access remote content using an internet connection.

see also cloud-based

remote working (*also* telecommuting)

Remote working is a way to do a job from home without needing to go to the workplace. You connect to the work computer using a comms link.

see also comms

removable disk

A removable disk is disk storage that you can take out of one computer and use in another.

rename

To rename a file is to give it a new name. In Windows, you can do this by right-clicking on the file icon and selecting from the menu.

WATCH OUT!

Changing the file extension can make the file unusable.

OK

repeater

A repeater is a piece of electronic equipment that picks up a signal and passes it on. This allows the signal to go further.

repeat loop

In some text-based programming languages, the keyword 'repeat' is used to define a condition-controlled loop,.

see also loop, while loop

replace *see* find

reset

To reset is to return settings to the values they started with. This may mean returning them to their default values.

see also default, settings

reply *see* comment

reserved word *see* keyword

resize

To resize something means to change its size. For example, you might resize an image.

see also window

resolution

The resolution of an image is the number of pixels in it. Images can be high-resolution or low-resolution.

see also high-resolution, low-resolution

restart

To restart something means to close a task down and start it again. Restarting a computer is the same as rebooting.

see also reboot

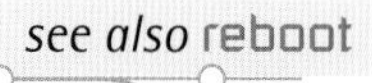

restore (*also* retrieve)

To restore something means to bring it back. If you accidentally delete a file, you can restore it from the recycle bin by right-clicking. If all your files are lost, you can restore from backup.

see also recycle bin, backup

restore settings

If you restore the settings of a device, you change them back to their previous values. You can use restore settings if you do not like the new settings.

see also settings, factory settings

result

The result of an operation is the value which is output at the end

see also operation

retinal scan *see* biometrics

retry

To retry means to try a task again if it has gone wrong. You usually make a change before you retry. For example, you might check that all devices are plugged in properly.

returned value

In programming, a returned value is a value made by a function. The returned value can be stored as a new variable. The command looks like this: `variable = function_name()`. The value returned by the function is stored in the variable.

see also function, variable

return key *see* enter key

retweet *see* tweet

RGB *see* colour

ribbon

Some software windows have an area at the top called the ribbon. The ribbon has icons on it that stand for different tasks and tools. You click on any icon to use that tool. The ribbon is a flexible display. It has different sections called tabs. By choosing different tabs you can see different collections of tools.

see also software window, tab, toolbar

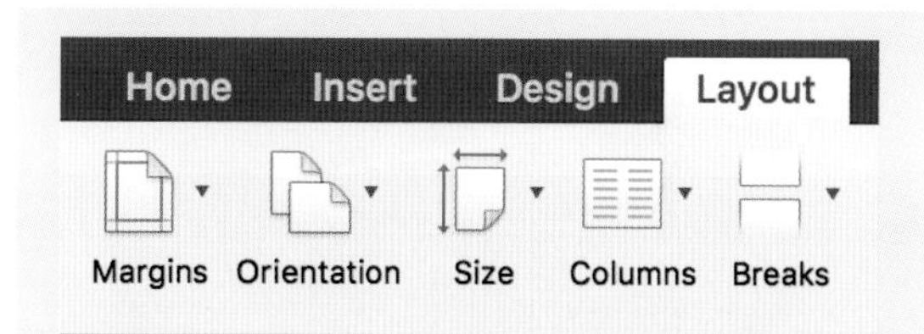

right-click *see* click

rip

To rip music means to copy it out of a file. For example, you can rip the music from a game or a video.

robot

A robot is a moving device that has a processor inside it. The processor controls the movement of the device. In stories, robots often look like people, though in real life they are more likely not to.

A B C D E F G H I J K L M N O P Q R S T U V W X Y Z

robust *see* code quality

ROM

ROM stands for *read-only memory.* Data stored in ROM cannot be changed. ROM is not volatile.

see also BIOS, memory, non-volatile RAM

Roomba

A Roomba is a kind of small robot with a vacuum cleaner attached. It rolls about the floor on its own and cleans the floor.

round brackets *see* brackets

router *see* network

routine *see* subroutine

row

A row is a horizontal group of cells. Spreadsheets and data tables are organised into columns and rows.

see also data table, spreadsheet, column

RPG

RPG stands for *role-playing game.* This is a game where you pretend to be a character in a story.

see also MMORPG

RSI

RSI stands for *Repetitive Strain Injury.* This is pain in the muscles of your wrist, hands and fingers caused by too much typing or by using the mouse too much.

> ⚠ WATCH OUT!
>
> Take regular breaks from the computer to avoid RSI.
>
> OK

RTF

RTF stands for *Rich Text Format.* This is a way of saving text files with some simple formatting. It was invented by Microsoft.

Ruby

Ruby is the name of a text-based programming language. It is suited to object-oriented programming (OOP).

see also high-level language, programming language, OOP

run

When you run a program the computer carries out the instructions in the program. This is also called executing the program.

see also execute

run speed

Run speed means how fast a computer does a task.

runtime error *see* error

Safari

Safari is the name of a web browser provided by the Apple corporation. You can only use Safari on Apple computers.

see also browser, Apple

Samsung

Samsung is the name of a Korean company that makes many different electronic devices such as smartphones and TVs.

sandbox game

A sandbox game is a computer game where you do not have to reach a goal. Instead, there are lots of features you can interact with. Fortnite, The Sims and Minecraft can be played as sandbox games.

sans serif font *see* font

satnav

Satnav is short for *satellite navigation.* It makes use of GPS to show you directions to a destination. You can get a satnav device for a car or you can use satnav on your phone.

see also GPS

save

To save a file is to copy it from electronic memory to permanent storage in a local storage device or in the cloud. This type of storage does not use electricity, so the file will not be lost when the power is turned off.

see also magnetic storage

› save as

A save menu might give you the choice 'Save As'. If you use this choice you can change how the file is stored. You might change the filename or the storage location or both.

see also file location

scalable

If a computer system is scalable it can be made to increase in size. For example, it might do a larger amount of work without any decrease in performance or speed. This is called rescaling.

› scalability

Scalability is the ability of a computer system to rescale.

a b c d e f g h i j k l m n o p q r s t u v w x y z

A B C D E F G H I J K L M N O P Q R S T U V W X Y Z

scam

A scam is a trick used to get personal details or money from people, for example by sending a trick email. Sometimes the scam includes identity theft.

see also phishing, identity theft, social engineering

scammer

A scammer is a person who does a scam.

scanner

A scanner is an input device that scans an image or an object. The features of the object are entered into the computer and are stored as digital data.

Scratch

Scratch is the name of a visual programming language. It is used to teach programming to young learners. Scratch commands control a sprite on the screen. You can write and run Scratch programs in a web browser.

see also high-level language, programming language, sprite, browser

screen

A screen is the flat area on a computer or smartphone that shows an image or data. It is a visual output device.

see also monitor, output device

screensaver

A screensaver is a moving image on the computer screen. It appears when the computer has not been used for some time. It shows that the computer has 'gone to sleep'. You can press any key to make the computer 'wake up'.

screenshot (*also* screen dump)

A screenshot is a picture of the screen. It is made when you press the print screen key. It is stored in the clipboard.

see also print screen, clipboard

script

A script is a short program that is typically written to do one task. Scripting languages are designed to be quick and simple to use. Scripts are often used to automate repetitive tasks, saving us time and effort.

see also automate

scroll

Scrolling is moving the pages of a document up or down on a screen. A document is often too big to see it all at the same time. Scrolling is like winding the document past the window. By scrolling through the document you can see it all.

see also document, mouse

scroll bar

The scroll bar is on the right of an open window. You move the little box (called the scroll box) up and down the bar to scroll through the document.

see also software window

scroll wheel

A mouse often includes a scroll wheel. You move the wheel with your finger. This makes the document on screen move up or down.

see also mouse

SCSI

SCSI stands for *Small Computer System Interface*. SCSI is a computer standard. It is a standard way to connect devices to the processor or motherboard. It mainly used to connect storage devices.

see also standard, port

search

To search means to try to find something you want, such as a piece of information. You can search a database for data or your hard drive for a file. You can search the internet for web content. Computer searching usually means finding a match to a search term. In a document, this is usually done using the 'find' function.

see also find, query

searchable

If a file is searchable, this means that the computer can search through the file to find a match. Text files are usually searchable. Image files are not searchable.

see also search, find

search engine

A search engine is a type of software on a website. A search term is entered into a search box and the software finds web pages that match the search. A search engine uses web crawlers to collect information about websites. Examples of search engines include Google and Bing.

> search results

The results of the search are shown as links to the web pages that match the search term.

> search term

A search term is a keyword that you enter into a search engine.

see also keyword

> search engine optimisation (*also* SEO)

SEO is a way of designing websites so that they come higher in the list of search results. This includes adding carefully chosen tags that help to get more visitors to the site.

see also tag

search algorithm

In programming, a search algorithm is a way to search for a value in a list or other dataset.

see also function

WORD BUILD

> binary search

A binary search is a way of finding a value in a sorted list or dataset. It only works if the list is sorted.

> linear search

A linear search is a way of finding a value in a list or dataset. It will work on any list whether it is sorted or not, but it is slower than a binary search.

see also binary, sort

search term *see* keyword, search engine

a b c d e f g h i j k l m n o p q r s t u v w x y z

A B C D E F G H I J K L M N O P Q R S T U V W X Y Z

secure browsing

There are some risks when you browse the internet. For example, some websites are set up to trick people. Secure browsing means using a web browser with features that help to avoid risks.

see also HTTP, encrypt

security

Computer security means all the ways we try to protect our data from loss and corruption. We also want to protect privacy.

see also corrupt, privacy, secure browsing, sensitive data

select

To select an object in a file means to pick it out so that the software can work with the object, for example to format or to copy it. You can select an image by clicking on it. You can select some text by dragging the mouse pointer over the text.

see also drag, object, click

selection structure *see* program structure

self-drive car

A self-drive car is a car that can steer past obstacles on its own. The human driver does not have to use controls. A number of companies are developing self-drive cars, including Tesla and Nissan.

semantic error *see* error

send

If you send data, it goes from one computer to another. You can also send a message or signal. Whatever you send travels down a comms link.

see also comms, email, transmit

sensitive data

Sensitive data is any data that people like to keep private, such as data about illness, bank accounts or relationships. There are strict laws about how sensitive data can be stored or shared.

see also personal data, data protection

sensor

A sensor is a type of input device. There are many kinds of sensor. A sensor measures something in the real world, such as temperature or light. The sensor converts the measurement to digital form and sends this data to the computer.

SEO *see* search engine

sequence *see* program structure

sequential access

(*also* serial access)

Sequential access describes a storage tape and means that the computer must wind through the tape in sequence to get to the stored data it needs.

see also random access, tape

serial

Serial describes events that happen one after the other. Serial processing is when instructions are executed one after the other. Serial transmission is when the bits of a signal are sent one after the other.

see also parallel

serial port

A serial port is a port that allows serial transmission.

see also port

serial transmission

Computer data is stored as bits. In serial transmission the bits are sent one after the other to another device. They can go down a single wire, or be sent wirelessly.

serif font *see* font

server

A server is a computer that 'serves' other computers with content. It is usually a large computer with lots of storage.

see also client-server

email server

An email server stores, receives and sends emails. The part of an email address after the @ symbol identifies the server.

network server

A network server sits in the middle of a client-server network.

web server

A web server holds internet content. It is connected to the internet and allows people on the internet to access the content.

server-side processing
see client-server

settings

Settings are adjustments to how a computer behaves. You can change the settings of your computer in lots of ways. For example, you can make the display bigger or you can add new software.

see also factory settings, restore settings

setup

Setup means getting your computer ready to use for the first time. You do this by plugging in hardware and connecting it, installing software and choosing settings. This can include new software, or a piece of hardware such as a new printer. In all cases it means getting it ready to use.

see also install, settings

share

A file or data can be shared. This means that more than one computer has access to it. This could be read-only or read-write access. Equipment such as a printer can also be shared.

see also access

shared resource (*also* network share)

A shared resource is any file or equipment that is shared using a network. Network users can have access to the resource. It helps with file sharing and teamwork.

see also file sharing

shareware

Shareware is a type of free software. You can use the software yourself or share it with others. There may be a limit to how much free use you can have. You can pay to go past the limit. Examples of shareware include business software, games and utilities.

shell

A shell is a simple user interface that makes it easier to use some software.

see also interface

shift key

The shift key is on the keyboard. It is marked with an up arrow. There are often two shift keys, one on the left, one on the right. You press the shift key at the same time as a character key. If you press a letter the shift key will make it upper case (capital letter). If you press one of the other keys the shift key will make the upper symbol on the key appear.

see also keyboard

shooter

A shooter is a computer game where the player pretends to shoot people or things, such as enemies or monsters.

see also first-person shooter

WORD BUILD

> shoot-'em-up

A shoot-'em-up is a type of shooter game. You are typically inside a vehicle or spaceship. You have to shoot large numbers of attackers, that keep coming.

shortcut

A shortcut is any method that gives you a quick and easy way to complete an action. For example, a shortcut icon is a copy of a file icon. You can put the shortcut in a convenient place such as the desktop of your computer.

show formatting

Formatting marks tell the computer how to format a document. In a WYSIWYG document the formatting marks are hidden. You click on the 'show format' icon to show (or hide) the marks, such as line breaks and tab marks.

see also format, WYSIWYG

silicon chip *see* microprocessor

Silicon Valley

Silicon Valley is a nickname for a place in California in the USA where lots of computer companies are based. Nowadays other places have similar nicknames, such as Silicon Fen in Cambridgeshire.

sim (*also* simulation)

Sim is short for *simulation*, which means an imitation of real life. A computer simulation is a model of a real-life event or system.

see also VR

Skype

Skype is the name of a web service that lets you make phone calls and video calls over an internet connection. It is cheaper than ordinary phone calls.

sleep mode *see* standby

slide

A slide is a screen display. Slides can have text, images, sound or video in them. A series of slides makes a presentation.

see also presentation, MS Office

slideshow *see* presentation

smart

Smart is a word that you can put before the name of an object to show that the object has a processor in it. For example, a smart fridge is one that has a processor inside and can connect to the Internet of Things.

see also Internet of Things, wearable technology

smartphone

A smartphone is a mobile phone with a processor inside. You can use the smartphone as a computer and it can run apps.

see also app

SMS

SMS stands for *Short Message Service*. This is the technical name for texting. We use SMS to send texts on our mobile phones. We use MMS to send text and images on our mobile phones.

see also text

snailmail

Snailmail means sending letters by the physical post system. This is much slower than computer communication such as email.

social engineering

Social engineering is a way to trick people on the internet. It is a type of scam in which someone tells false stories to trick people into sending money or giving up personal data and passwords.

see also identity theft, hacking

social media

Social media means all the websites and apps that people use to communicate with each other on the internet. The content of social media sites is made by the users. Facebook and Twitter are examples of social media.

see also Web 2.0, social network, Facebook, Twitter

social network

Social networks are made of human friendships and relationships. On the internet, a social network is any website or app that allows people to be social. You interact with other people through the internet.

see also social media

socket *see* port

software

Software means a program. A program is a set of instructions that a computer can run. The computer will carry out the instructions in the program. There are two main types of software: application software and system software.

see also hardware

application software

Application software performs a useful task for the person who uses it.

see also app

system software

System software controls the hardware and workings of a computer, tablet or phone.

software engineering

Software engineering means making software. Most of the work of a software engineer is designing or writing programs.

see also program

software window

On a modern computer system most apps open in a window, called a software window. A software window has some standard features. The exact features are different on different computer systems. Open software windows are shown as icons on the task bar.

see also desktop, window

solid-state storage *see* flash memory

a b c d e f g h i j k l m n o p q r s t u v w x y z

A B C D E F G H I J K L M N O P Q R S T U V W X Y Z

sort

1 To sort data is to arrange it in a particular order, such as numerical or alphabetical order.

2 If you do a sort you put data in a particular order, such as numerical or alphabetical order.

sort function

A sort function is a function that sorts a data structure. The elements of the structure are put in order, such as numerical or alphabetical order.

sort order

Sort order means the method used to do a sort. Number values are put in numerical order. Character values are put in alphabetical order.

sound card

A sound card is an expansion card that makes the computer better at processing sounds.

source code

see code

space bar

The space bar is at the bottom of a keyboard. You press it to make a space between characters.

see also keyboard

spam (*also* junk mail)

Spam is unwanted email. Computers make it cheap and easy to send millions of emails. The emails may be adverts or they may be scams. They can fill up your email system with junk (unwanted emails).

spam folder

Anti-spam software can spot spam emails. They are sent to a special folder on your email system called the spam folder.

speaker

A speaker is an output device that sounds come out of.

speech recognition

(*also* voice recognition)

Speech recognition is a computer feature that makes the computer able to understand the sound of a human voice. If you say a word the computer will return it as text, as if you typed in the text.

see also natural language interface, voice-activated

speech synthesiser

A speech synthesiser is a type of computer system that is able to read text files and read them out loud like a human voice.

spellcheck (*also* spellchecker, spell-checker)

A spellcheck is a feature of word-processing software. The computer has a list of words. If you type something that is not on the list, the computer marks it as a spelling mistake. You can right-click to show a list of possible correct spellings.

spider *see* web crawler

split screen

A split screen is a screen on which two or more windows are open. You can see both open windows at the same time.

see also window

spoofing

Spoofing means pretending to be someone else as a kind of scam. For example, trick software can disguise an email address or a URL to look like someone else.

see also scam, identity theft, hacking, phishing, spam

Spotify

Spotify is the name of an internet service that lets you listen to music over an internet connection.

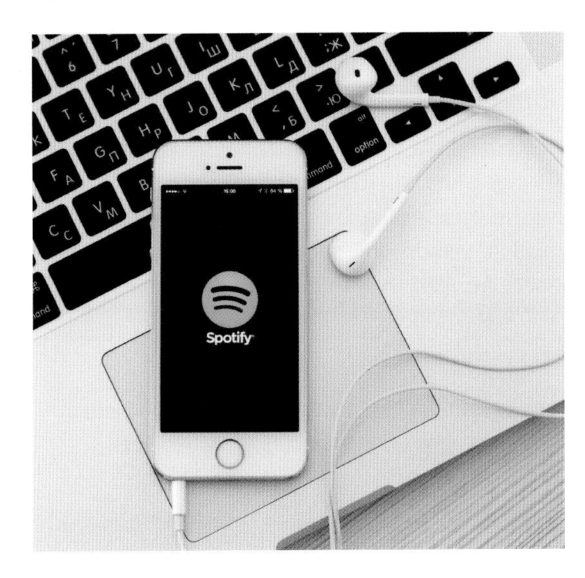

spreadsheet

A spreadsheet is a way of setting out data in a table of rows and columns. The spreadsheet has tools to process the data. In particular, it can do calculations. It does this by using formulas with cell references. There are different apps you can use for creating spreadsheets, such as Excel.

see also formula, cell reference

sprite

A sprite is a little picture on the screen. The commands in a program make the sprite move about on the screen.

see also Scratch

spyware

Spyware is software that hides on your computer and records everything that you do. Some spyware is malware that hackers use to find out your password. Some companies use spyware to find out what their employees are doing.

see also malware, privacy

SQL

SQL stands for *Structured Query Language*. SQL is a programming language that is used to write database queries. SQL can be pronounced 'S-Q-L' or 'sequel'.

see also database, relational database

square brackets *see* brackets

SSD

SSD stands for *Solid State Drive*. This is an alternative to a hard disk drive. It uses flash storage.

see also hard disk, flash memory

standalone (*also* stand-alone)

A standalone system is one that works on its own. It does not need to be connected to any other hardware.

standard

A standard is a shared way of doing something. For example, a USB is a standard way to link a device to a computer. Shared standards let computers communicate. If they all used different standards they would not be able to link together.

see also protocol

standby (*also* sleep mode)

A device that is on standby is using very little power but has not been completely shut down. It is quick and easy to 'wake it up' again.

start

To start software means to begin to run it. The computer carries out the instructions in the software.

see also start up

a b c d e f g h i j k l m n o p q r s t u v w x y z

A B C D E F G H I J K L M N O P Q R S T U V W X Y Z

Start button

The Start button is a feature of Windows computers. It is at the lower left corner of the computer screen. If you click the Start button you will see the Start menu.

see also Start menu

Start menu

The Start menu is a feature of Windows computers. It is a list of all the software available on the computer. From the Start menu you can start any software you want.

see also menu, Windows

start up

You can start up a computer system or a piece of software. Starting up a computer system is the same as booting up. Power goes into the system and it gets ready for use. Starting up software means loading it into RAM. The software is now ready to use.

see also boot, RAM

statement

In programming, a statement is another word for a command. The word statement is often used for a command that takes several lines of code, such as a loop.

see also command

storage

Storage holds digital data in a permanent form without using electricity. Data in storage is not lost when the electricity is turned off. There are many different types of storage, including optical storage, magnetic storage and flash storage.

see also magnetic storage, optical storage, flash memory, memory, volatile, non-volatile

storage device

A storage device is any device used to store data. The data is stored without electricity, so that it remains after the computer is switched off.

storage location

The storage location of a file is the drive and folder where it is saved.

see also drive, folder, location

store *see* storage

storage tape *see* tape

streaming

Streaming is a way to look at internet content such as music and video. You can look at the first part of the stream while the rest is downloading. It is held in memory and does not use up your storage space. For example, YouTube uses streaming to let you watch videos.

› live streaming

Live streaming means you can show the videos 'live' so viewers can watch as the video gets recorded.

see also download

string *see* data type

strong password *see* password

stylus

A stylus is an input device that you use with a touchscreen. It controls the screen more accurately than a finger.

see also touchscreen light pen

subdirectory
see folder

subfolder *see* folder

subheading *see* heading

submit button

An online form often has a submit button. After you have entered all the data you then click the submit button. This will send the information.

see also online form

subroutine (*also* sub-procedure, subprogram, routine)

In programming, a subroutine is a named block of code. The programmer writes the subroutine and gives it a name. There are two main types of subroutine: procedures and functions.

see also procedure, function

sum

Sum is sometimes used to mean any calculation. It can also be used more precisely to mean the result of adding together a series of numbers.

see also AutoSum

sunrise industry

A sunrise industry is a type of industry based on new technology. As the technology develops, the industry should get more successful but it can be hard to predict which industries will do best.

supercomputer

A supercomputer is a very powerful computer. It is a computer that can work very fast and can process lots of data. Supercomputers are used in many types of scientific investigation, most famously in weather forecasting and the Large Hadron Collider at CERN.

see also mainframe

support (*also* user support)

Support means help for software users. Support might include a help desk, training or a user manual.

see also help, help desk

surf

Surfing the internet is a way of browsing from website to website by clicking on hyperlinks.

see also browse, hyperlink

switch *see* network

syntax error *see* error

system administrator (*also* SysAdmin)

A system administrator is the person in charge of a computer system such as a network.

see also admin privilege

system software *see* software

A B C D E F G H I J K L M N O P Q R S T U V W X Y Z

T *see* order of magnitude

tab

1 (*also* **tab key, tab stop**) Tabs are fixed positions within one line of a document. The keyboard has a 'tab' key. If you press this key the cursor moves to the next tab position. You can use tabs to make sure text is lined up neatly on the page.

see also keyboard

2 A tab is a feature of some software windows. A tab holds a group of related icons. You can click on a tab to see the tools you need.

see also ribbon

tablet

A tablet is a small portable computer with a touchscreen. Almost all actions use the touchscreen.

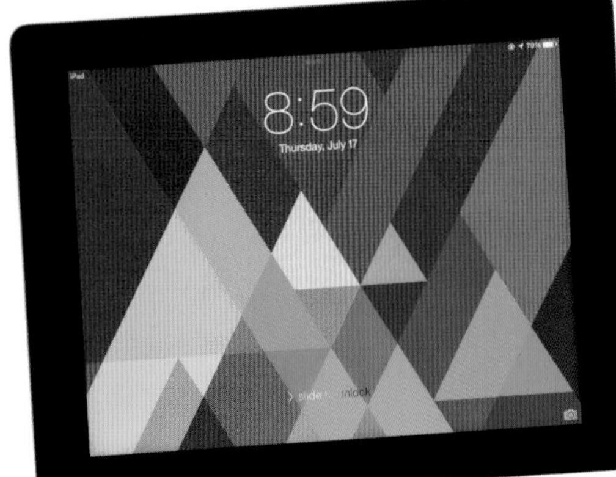

tag

A tag is a piece of metadata that you add to a file or web page. It describes the data.

see also metadata

tape (*also* storage tape)

Magnetic tape is used to store data. The computer must wind through the tape to get the data it needs (sequential access). This means that access is slow. However, tape is cheap storage and it is often used for backups.

see also sequential access, backup

task

Task is a general word for an action completed by the computer.

taskbar *see* desktop

task manager

The task manager is a piece of software within a computer's operating system. It keeps track of all the different tasks that the computer is doing at that time.

see also operating system, multitasking

TCP/IP

TCP/IP is the term used for a set of protocols that help to make the internet work. It includes the two separate protocols TCP and IP.

see also protocol

WORD BUILD

> TCP

TCP stands for *Transmission Control Protocol*. This controls how the computer sends data down an internet connection. It uses packet-switching.

see also packet-switching

> IP

IP stands for *Internet Protocol*. This makes sure internet signals go to the right address.

see also IP address

technical support

Technical support means help with hardware or software. Most organisations have technical support. If a piece of equipment or software is not working, you can contact technical support.

technology

Technology is a general word for human inventions that help us to do things. Computers are a key example of modern technology. Often when people talk about technology they mean computers.

telecommunications (*also* telecomms)

Telecommunications means long-distance communication links. We often shorten the word to telecomms. Tele- is a prefix meaning 'long-distance'.

telecommuting (*also* teleworking, remote working)

teleconference (*also* conference call)

A teleconference is a phone call shared between more than two people.

see also videoconference

template

A template is an outline of a document that you can use to structure the document. Documents that are made with the same template will look alike.

tera (*also* T) *see* order of magnitude

terminal (*also* workstation)

A terminal is a computer or workstation connected to a larger computer. It lets a person use a large computer from a distance. A terminal has input and output devices such as screen and keyboard. In a more general sense, a terminal is the end of a connection.

see also mainframe

dumb terminal

A dumb terminal is one that cannot do any processing. It only sends and receives data from the remote computer.

intelligent terminal

An intelligent terminal is one that has a processor. It can do some processing itself.

testing

Testing means trying out a program to find errors. The program should be tested with lots of different inputs. If the testing finds any errors, they must be fixed before the program is used.

see also debug, error, beta test

text

1 Text is content made of characters (symbols such as letters, digits and punctuation marks). Text is saved to a file encoded in character sets like ASCII or Unicode.

see also ASCII, Unicode, data type

2 (*also* text message) A text is a short written message sent between mobile phones using a technology called SMS.

see also SMS

3 To text is to send a text message.

text-based language

A text-based language is a type of programming language. The commands are written as lines of text.

see also visual programming language

text box

A text box is any box that can hold text. Text boxes are used in two ways. They can show text to the user, for example in websites or presentations. They can also be used to collect text from the user. The user types text into an empty box, for example in an online form.

see also online form

text file

A text file is a file made of text characters and nothing else. It often has the file extension .txt.

text formatting

Text formatting is a way to change the appearance of text, for example its font, size or colour.

font colour size
font colour size
font colour size

text size *see* font

text-to-speech

Text-to-speech is a computer system that is able to read text files and read them out loud like a human voice.

see also speech synthesiser, speech recognition

thumbnail

A thumbnail is a small image of something on a computer screen. It is about as small as your thumbnail. A thumbnail shows you what a larger image looks like. This lets you review lots of images on the screen. An example is the photos in a photo album. You click on a thumbnail to open the image.

see also icon

thumb print *see* biometrics

tick box *see* checkbox

time out

If a comms link times out, it turns itself off automatically after it has not been used for a certain length of time.

title bar

The title bar appears at the top of an open window. It shows the name of the software app running in the window.

see also software window

TLD

TLD stands for *Top Level Domain*. The TLD is the last part of a website address, after the final dot. It tells you the type of website. Examples of TLDs are: **.com** (a business site); **.uk** (a site from the UK); **.gov** (a government website).

toggle

A toggle is a single button that does two things. If a feature is OFF it turns it ON. If a feature is

ON it turns it OFF. Caps Lock is an example of a toggle.

token

A token is a single element of a programming language. A programmer makes a program by fitting different tokens together. Examples of tokens are values, and expressions which represent values. Tokens which can change values include operators, program keywords and predefined functions. Tokens made by the programmer include identifiers (names) of variables and subroutines.

see also value, expression, operator, keyword, predefined function, identifier, variable, subroutine

toner

Toner is a fine dust that is used like ink in a laser printer.

see also printer

tool

Tools are icons on the toolbar or ribbon. Each tool stands for a feature offered by the software. You click on the tool to use that feature.

see also toolbar, ribbon

edit tools

Edit tools let you cut, copy and paste objects (pieces of text, images and so on).

format tools

Format tools let you change the appearance of objects (pieces of text, images and so on), for example the colour of text.

toolbar

Some software windows have an area at the top called a toolbar. The toolbar has icons on it that stand for different tasks and tools. You click on any icon to use that tool. A toolbar is a fixed display. It always shows the same tools.

see also software window, ribbon

touchpad

A touchpad is an input device that you find on many laptops. It is a pointer device. You can use it instead of a mouse.

touchscreen (*also* touch screen)

A touchscreen is a type of screen, found on devices such as tablets and smartphones. A touchscreen can be used to input instructions and choices to the computer. When you touch the screen, a signal is sent to the computer. The computer knows what part of the screen you touched. This is an alternative to using a keyboard or a mouse.

see also tablet, smartphone, screen, pointer device

tower computer

A tower computer is a type of desktop computer that is tall in shape. It is placed next to the monitor or on the floor.

tracker ball (*also* trackball)

A tracker ball is an input device. It is an alternative to a mouse. The device stays still on the desk and you move a ball on the device with your fingers.

see also mouse

transaction

1 In everyday use, a transaction is an event in which things are exchanged or sold. Buying goods online is a common example of a transaction.

2 In computing, a transaction is an event in which data is changed. A transaction might change several items of data. But a transaction is always handled as a single unit. There is no such thing as half a transaction. Either all items of data are changed or none.

transaction processing

Transaction processing is a way that some computers work. The computer gets all the information about a transaction and then it carries out the transaction in one task.

see also transaction

translate

In programming, to translate means to turn source code into machine code. Source code is written by the programmer. It has to be turned into machine code so that the computer can understand it. After the code has been translated, the computer can run (execute) the code.

see also code

translator

In programming, a translator is software that translates source code into machine code. Two types of translator are a compiler and an interpreter.

see also compiler, interpreter

transmit

If you transmit a signal, message or some data, you send it from one computer to another.

> transmission

A transmission is a signal or message that is transmitted. It can also mean the act of sending the signal or message.

trigger

A trigger causes an event to occur or is caused as a result of an event. Triggers are used in event-driven programming languages. A trigger makes the computer carry out a command. An example of a trigger would be the user clicking on a button on the screen.

see also event-driven programming

trojan *see* malware

troubleshooting

Troubleshooting means finding and fixing problems or faults.

> troubleshooter

A troubleshooter is a piece of software that helps find problems in code.

truth table

A truth table is a grid that describes a logic circuit or a logic gate. It sets out all the possible inputs and outputs.

see also logic gate, logic circuit

tuple

In programming, a tuple is a data structure that is fixed in size like an array. The values in a tuple cannot change.

see also data structure

Turing test

A Turing test is an imaginary test. If you are communicating through a computer link, can you tell the difference between talking to a computer and a person? There is a disagreement about whether you can always tell the difference or not. The Turing Test is named after Alan Turing, who was a pioneer of computing.

turtle (*also* floor turtle)

A turtle is a little robot on wheels that runs about on the floor and can draw with a pen. It is controlled by a program. Controlling a turtle is a fun way to learn programming.

turtle graphics

Turtle graphics is an on-screen display that acts like a turtle robot. This lets you do turtle programming even if you do not have a turtle robot.

tweet

1 To tweet means to post a message on the social network site Twitter. All your followers can read your message.

see also Twitter, follow

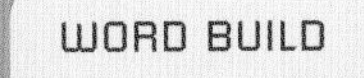

2 A tweet is a message posted on the social network site Twitter.

WORD BUILD

> retweet

To retweet means to send someone else's tweet to your followers on Twitter.

Twitter

Twitter is the name of a social network site on which you can post messages and read other people's messages. The messages are limited to 280 characters. You can attach links and images.

see also social network

typeface *see* font

UAV *see* drone

underline

Underline is a way to emphasise text in a document by showing a line underneath it.

undo

To undo an action means to reverse it. You can choose 'undo' if you have made a mistake. There is often an undo button at the top left of a window. Ctrl-Z is the keyboard shortcut.

see also keyboard shortcut, software window

Unicode

Unicode is a way to encode text characters. It is an extension of ASCII. Unicode has thousands of codes. It has codes for all the alphabets and character sets in the world and for other symbols such as emojis.

see also ASCII

Unicode consortium

The Unicode Consortium is a non-profit organisation that agrees the Unicode number for every character. This makes sure that the codes are standardised so that computers can communicate with one another.

A B C D E F G H I J K L M N O P Q R S T U V W X Y Z

uninstall

To uninstall software is to remove it from computer or phone. It frees up storage space, but you cannot use the software any more.

see also install

unzip

To unzip a file means to remove file compression and so return it to its original, larger size.

see also zip, compress

update

1 To update something is to bring it up to date. You may get a message from the manufacturer asking if you want to update software. If you say yes it will happen automatically.

2 An update is a change made by updating something.

upgrade

To upgrade is to make something better, usually by adding extra features. You can upgrade software over an internet connection.

upload

1 To upload a file means to put it on a web server. Other people are then able to access the file over an internet connection.

see also server, download

2 An upload is a copy of a file that you upload to a web server.

upper case *see* case

URL

URL stands for *Uniform Resource Locator*. The URL is the address of a web page. It is made of letters and other characters. It sets out the path to get to the website, then the path on the website to that particular page.

see also IP address, web address

usability

Usability means how easy it is to use software. A GUI (Graphical User Interface) makes software easier to use. So do clear messages on screen.

see also GUI

USB

USB stands for *Universal Serial Bus*. USB is a standard for joining devices together, for example to send data from one device to another. Many devices use USB connections. Electricity can be sent through a USB connection, to recharge the battery of a device such as a phone.

see also standard, port

user (*also* end user)

A user is anybody who uses the hardware and software of a computer system or other device. The user provides inputs and receives outputs.

see also input, output

user account

A user account is a way of keeping track of computer users. This is important in a shared system used by many people, such as a network or a social media site. The user account stores the username, password and other important information. When you log in you connect to your account. Then you can access your files or entries.

see also log in, log on, account

user error *see* error

user-friendly

User-friendly is a way to describe software. Software that is user-friendly is easy to use. It has a good user interface.

see also user interface

user group

A user group is a group of people who use the same software. A user group often meets online to share tips about how to use the software.

user interface

The user interface is what a user sees when they use software. It includes messages to the user and ways for the user to enter input.

see also user, interface

username

A username is your identity when you use an online service. It might not be your own name.

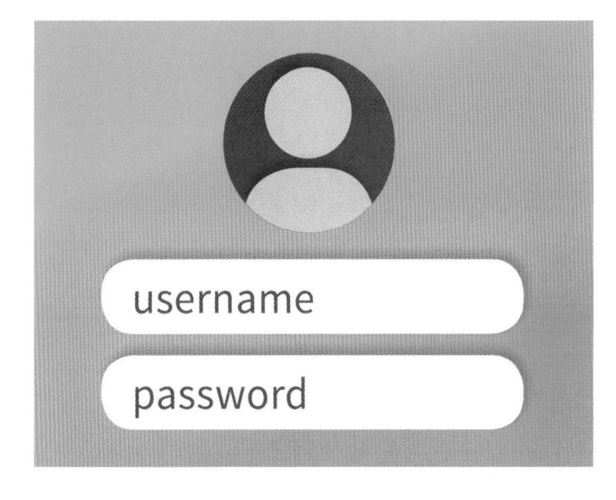

user support *see* support

utility

A utility is an extra piece of system software that is not part of the operating system. An example of a utility is antivirus software.

see also antivirus software

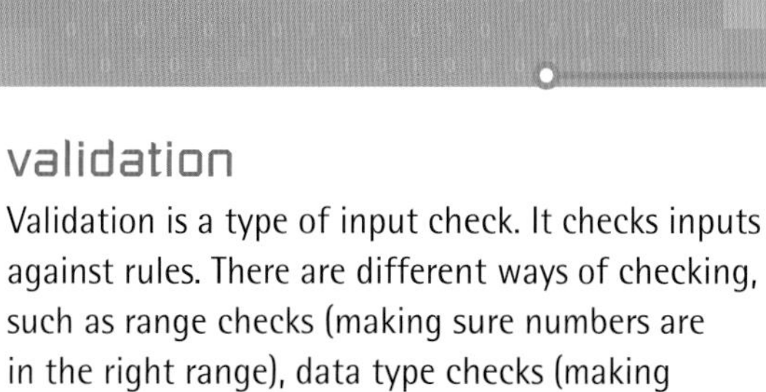

validation

Validation is a type of input check. It checks inputs against rules. There are different ways of checking, such as range checks (making sure numbers are in the right range), data type checks (making sure data is the expected type) and list validation (checking input against a list of allowed inputs).

see also input check

value

In programming, a value is one item of data. Values can be of different data types. Here are some examples: True (a Boolean value); 99 (a numerical value); "hello" (a string or text value).

see also data, data type, expression

vapourware

Vapourware means software that is planned to be produced, but that you cannot get yet. Sometimes software is announced but it never appears at all.

see also software

variable

A variable is an area in computer memory. A programmer can give the memory area a name. This is called declaring a variable. A programmer can store a value in the named area. This is called assigning a value to a variable. The value stored can vary, which is why it is called a variable.

see also declare, assign, initialise, constant

VDU

VDU stands for *Visual Display Unit*. It is another name for a monitor or screen.

see also monitor, screen

vector graphics

Vector graphics is a way of storing an image in digital form. The computer stores the length and angle of the lines and shapes that make the image. It fills the spaces with colours and patterns.

see also bitmap graphics

> **WORD BUILD**
>
> **vector image**
>
> A vector image is an image made using vector graphics.

verification

Verification is a type of data check. The data is entered twice. The two entries should match. For example passwords are often verified when you change them online to make sure you have entered them correctly.

see also input check, authenticate

version

A version of a piece of software is one of the different forms it appears in. Later versions often have better features. The version may be shown as a number after the software name, for example Windows 7/8/10 or MacOS 10.1/10.2/10.3. If you update software you now have a later version.

see also software, update

video card *see* graphics card

videoconference

A videoconference is a meeting in which people in different places can see and talk to each other using video screens.

video game *see* game

viral marketing

Viral marketing is a way for a business to promote or advertise its products by encouraging its customers to share information about them on social media.

see also social media

virtual keyboard

A virtual keyboard is a keyboard on a touchscreen that you can use instead of a physical keyboard. You type the keys by touching the keyboard on the screen.

virtual memory

Sometimes the computer's electronic memory is full. Then the computer has to use non-electronic storage as extra memory, for example the hard disk. This extra memory is known as virtual memory. Using virtual memory makes the computer go more slowly.

virtual reality *see* VR

virus (*also* computer virus)

A virus is a type of software that copies itself into other files. It usually has a damaging effect on the data in the files. We say that the files are 'infected' with the virus. You can use antivirus software to get rid of viruses from your computer.

see also malware, antivirus

visual programming language (*also* block-based language)

A visual programming language is a type of programming language. The programmer makes a program out of blocks that fit together. Scratch and App Inventor are visual programming languages.

see also Scratch, app, programming language, text-based language

voice-activated

Voice-activated software is software that you can control by giving voice commands. The software uses speech recognition.

see also speech recognition, natural language interface

voice recognition *see* speech recognition

volatile

Volatile describes computer memory that uses electricity. When the electricity is turned off, the data is lost. To keep the data it must be copied to non-volatile storage.

see also non-volatile

VR

VR stands for *virtual reality*. This is an experience (called a simulation) made by a computer but similar to real life. It might be 3D (3-dimensional). People most commonly experience VR by wearing a special headset.

see also sim, AR

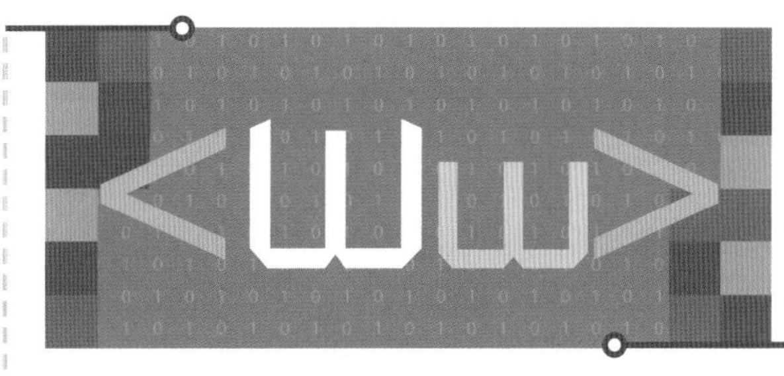

wallpaper *see* desktop

WAN

WAN stands for *Wide Area Network*. A WAN is a network where the computers are not close together. They are often in different cities or even different countries. The internet is a WAN.

see also LAN, network, internet

WAP

WAP stands for *Wireless Access Point*. A WAP lets wireless devices connect to a wired network.

see also hotspot

WAV

WAV is a file format used for storing sounds. It is higher quality than MP3, but the files are larger. WAV is pronounced 'wave'.

see also MP3

weak password *see* password

wearable technology (*also* wearable tech)

Wearable technology means electronic devices that you can easily wear or carry on your body. A device might be sewn into clothing. It might be something you can wear like a wristwatch. It might even be implanted under your skin.

see also fitness tracker, smart, Internet of Things

a b c d e f g h i j k l m n o p q r s t u v w x y z

A B C D E F G H I J K L M N O P Q R S T U V W X Y Z

Web *see* world wide web

Web 2.0

Web 2.0 means the modern kind of website that has interactive content. Users help to make the content.

Web 3.0

People expect a new type of website will be invented one day, which they call Web 3.0. However, nobody can agree exactly what it will be.

web address (*also* URL)

Every web page on the internet has its own address. The web address tells the web browser how to find the web page. Another name for a web address is a URL.

see also URL, web page, IP address

web browser *see* browser

webcam

A webcam is a camera joined to an internet connection. It may be part of bigger device such as a phone. It lets you see pictures or video over the internet. Webcams are used for video phonecalls, such as Skype calls.

web crawler (*also* spider)

A web crawler is a piece of software used by a search engine. The web crawler follows web links and records what is on each web page. This is part of how search engines can find the right page.

see also search engine, bot

web hosting

Web hosting means holding web pages and other internet content on an internet server.

see also internet server

web log *see* blog

web page

A web page is a document written in HTML (HyperText Markup Language). When you look at a web page, the HTML comes down an internet connection to your computer so that you are able to see the web page in your browser.

see also HTML, browser

web portal

A web portal is a website that brings together information and useful links on a certain topic. It presents the information in a useful way.

see also website

web robot *see* bot

web server *see* server

website

A website is a collection of web pages. They are all hosted on the same server.

WhatsApp

WhatsApp is the name of software that lets you send text messages over an internet connection. It is owned by Facebook.

while loop

In many text-based programming languages, the keyword 'while' is used to make a condition-controlled loop.

see also keyword, loop, repeat loop

```
password = ""
tries = 0
while password != "sesame":
    password = input("enter the password: ")
    tries = tries + 1
```

Wi-Fi (*also* wi-fi, Wifi, wifi)

Wi-Fi is a shared standard for wireless communications. If a device is Wi-Fi enabled it can use Wi-Fi links. There are lots of Wi-Fi links, so this is very useful.

see also protocol, standard, ethernet

Wi-Fi hotspot (*also* hotspot)

A Wi-Fi hotspot is a place where a device can connect to a Wi-Fi network.

see also hotspot

wiki

A wiki is a type of website. The users of the website can change what is on the site. They can edit it using a web browser.

Wikipedia

Wikipedia is the name of a free online encyclopedia. It is an example of a wiki. The users create the content.

see also wiki

⚠ WATCH OUT!

Be aware that the content of a particular Wikipedia article may not always be completely accurate. However, used with care it can be a useful resource.

OK

wildcard (*also* wild card)

A wildcard can be used in a search or query. It shows a place in the query where any value is allowed. The asterisk * is often used as a wildcard. For example, if you search for M* this will find any word that begins with M or contains 'm'.

see also search, query

window

A window is an area of a computer screen. Content is displayed inside the window. Most modern GUIs (Graphical User Interfaces) use windows. Two types of window are a dialogue box and a software window.

see also dialogue box, software window, GUI

WORD BUILD

> active window

You can have several windows open at once, but only one is active. The active window is at the front of the screen display. Your typing or other actions affect the active window.

> maximise

If you maximise a window it fills the whole screen except the task bar.

> minimise

If a window is minimised it shrinks to just an icon on the task bar.

> move

If a window is not full size you can move it. You move a window by dragging the title bar of the window.

> resize

You can drag the border of the window in order to change the size of the window. You can also use the resize buttons to make a window bigger or smaller.

see also drag

Windows (*also* MS Windows)

Windows is the name of an operating system made by Microsoft. It may be used on many different types of computer.

see also operating system

WinZip

WinZip is the name of an application that compresses and decompresses files.

see also compress, decompress, zip

wireless

Wireless describes something that works without a wire connection. The signal may be sent by radio waves, for example. Input and output devices can be wireless, such as a wireless mouse. Networks can also be wireless.

wireless network

A wireless network is one where the communication links are made without cables, for example by using radio waves.

see also network Wi-Fi

wizard

A wizard is a user-friendly way for a piece of software to prompt a user for several pieces of information. For example, a wizard might help you to change your computer's display settings.

Word

Word is the name of a word-processing program. It is made by Microsoft and is part of MS Office.

see also MS Office

word processor

A word processor is a software application that lets you make and edit documents.

> word processing

Word processing is the use of a computer to make and edit documents.

word size

Word size is a way of measuring the power of a processor. Word size is measured in bits. It tells

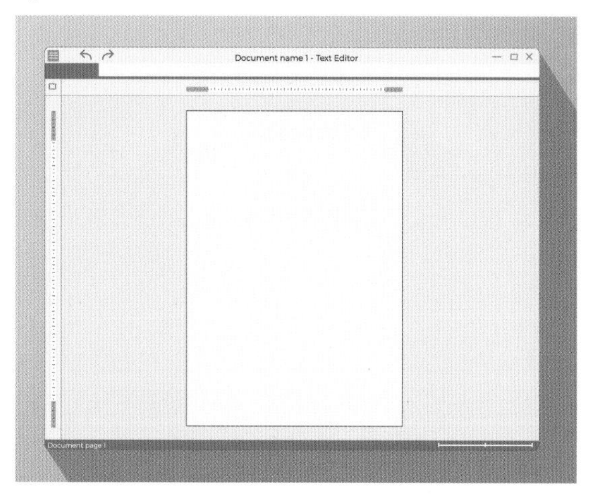

you how many bits the processor can change in one operation. Bigger word size generally means a faster processor.

see also 32-bit, 64-bit

workstation

1 A workstation is a place where a user can work with a computer.

2 Your computer at school or work can also be called a workstation. It might be a terminal connected to a mainframe or it might be a powerful standalone computer.

see also terminal, standalone

world wide web (*also* www, the Web)

The world wide web is made up of all the web pages in the world. They are connected through the internet.

see also internet

worm *see* malware

write

When a computer writes data, it puts new data into a file in storage. This might involve adding new data or deleting it.

see also read

write-protected

If a file or storage area is write-protected, the computer cannot write to the storage. This means that none of the data can be changed or removed.

see also access, write

WYSIWYG

WYSIWYG stands for *What You See Is What You Get*. It is pronounced 'wizzy-wig'. This means that what you see on the screen is what you get, in that exact format, when you print out the document. The screen is an accurate picture of the document.

www *see* world wide web

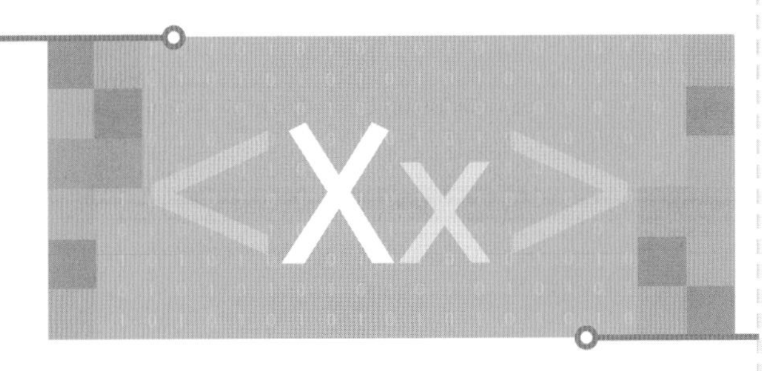

x coordinate *see* coordinates

XML (*also* Extensible Markup Language)

XML stands for *Extensible Markup Language*. XML is a language used to make documents that both computers and people can read.

see also HTML

y coordinate *see* coordinates

YouTube

YouTube is the name of a website that hosts videos. It holds millions of videos.

see also streaming

zip

To zip a file means to compress it (make it smaller). A large file is compressed so that it takes up less storage space.

see also compress, unzip, WinZip

zoom

To zoom in on an image means to make it bigger on the screen. The amount of zooming in is often shown as a percentage. For example, 200% means twice as big. The zoom feature can also be used to make an image smaller.

a b c d e f g h i j k l m n o p q r s t u v w x y z

The Computing Timeline

1822
Charles Babbage designs a machine to do automatic calculations. He calls it the Difference Engine. It works not by electricity but by cranking a handle to turn gear wheels. His friend Ada Lovelace invents programs that could run on the machine.

1939
Alan Turing starts work at Bletchley Park. His task is to break the secret codes used by the German high command to send wartime messages to its armies. To do this he builds an electronic machine called the Bombe. It is very similar to a modern computer. His work helps the Allies win the Second World War.

1950–51
The first computer with a program stored in its memory is made in the USA. It is called UNIVAC 1101. The first programming language is made by Grace Hopper. It turns commands written by a programmer into the 1s and 0s of machine code.

1961
The microprocessor is invented by Jack Kilby and Robert Noyce. An entire computer processor can now be made from a single chip of silicon. The age of the microcomputer has begun.

1969
A new network is set up to link military computers in the USA. It is called ARPANET. The network soon includes non-military computers. The method invented to link the computers is the basis of the modern internet.

Charles Babbage
1791–1871

Mathematician, philosopher, inventor, mechanical engineer

- 1822: designed the first mechanical computer called the Difference Engine
- 1837: designed a second computer called the Analytical Engine
- called the 'father of the computer', has his work on display in the Science Museum in London

Ada Lovelace
1815–1852

Mathematician and writer, daughter of the poet Lord Byron

- 1843: designed the first algorithm for the Analytical Engine designed by Charles Babbage
- considered the first ever computer programmer
- had the vision that computers could go far beyond calculating and number crunching

Alan Turing
1912–1954

Mathematician and computer scientist

- 1940: designed the Bombe while working in Bletchley Park, the code-breaking centre, during the Second World War
- The Bombe: an early computer that broke German communication codes and helped the Allies win the war
- helped design early computers at Manchester University and designed ACE, one of the first stored-program computers
- developed the Turing Test, an attempt to define intelligence in computers

1972

The first computer game console, the Magnavox Odyssey, goes on sale. It has one game—a table tennis game called Pong. By 1975, 350,000 consoles have been sold worldwide.

1975

Steve Jobs and Steve Wozniak sell the first Apple computer. For the first time programmable computers are affordable and available to ordinary people in their own homes.

1981

The first laptop computer goes on sale. It is called the Osborne 1. It is expensive and heavy with a tiny screen, but at their peak, sales are 10,000 a month.

1989

Tim Berners-Lee launches the World Wide Web by inventing the idea of documents that are readable over an internet connection.

1995

Jeff Bezos sets up a website on the World Wide Web. He decides to sell books. He calls the website Amazon. It grows into a massive international business selling all types of goods online.

2003

A student called Mark Zuckerberg sets up a website called Facebook for students at his university. In 2006 he opens it up to anybody to use. By 2019 Facebook has over one billion users.

Bill Gates
1955–

Business magnate, investor, author and philanthropist

- 1975: co-founded Microsoft with Paul Allen, which eventually became the world's largest PC software company
- Windows 10: runs on about 85% of personal home computers today and Microsoft is worth about $850 billion
- is the world's richest man after Jeff Bezos, the founder of Amazon
- 2014: left Microsoft to focus on his charity work to improve global health

Sir Tim Berners-Lee
1955–

Engineer and computer scientist, best known as the inventor of the World Wide Web

- 1989: invented a protocol called HTTP to download a page of information from the internet to your computer
- also invented HTML, the language in which all web pages are written
- founded the W3C (World Wide Web Consortium), the governing body of the web
- 2004: received a knighthood
- 2016: received the Turing award for inventing the World Wide Web

Shigeru Miyamoto
1952–

Video game designer and producer for game company Nintendo

- 1979: joined Nintendo and created some of the best known games in history
- his most iconic video game characters: Super Mario Bros, Donkey Kong and Zelda
- is the Representative Director and Creative Fellow of Nintendo

The Computer and its Devices

Every computer has a **processor** inside it. A processor uses electronic signals to do its work. It processes **data** to create new **information**.

Monitor

Tower computer

Processor is in here.

Hard disk or hard drive is in here.

Ports

Input devices let users input the instructions and data that make the computer work. For example:

- **Keyboard**: text input
- **Mouse**: positional input (point and click)
- **Microphone**: voice instructions

Keyboard

Mouse

Output devices let the computer provide results in a suitable form for the human user. For example:

- **Monitor**: visual output
- **Speaker**: sound output
- **Printer**: output on paper (hard copy)

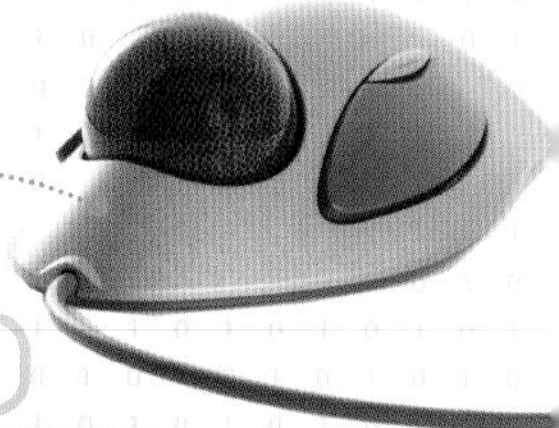

Tracker ball

This device does not move on the desktop. The tracker ball rotates to move the pointer.

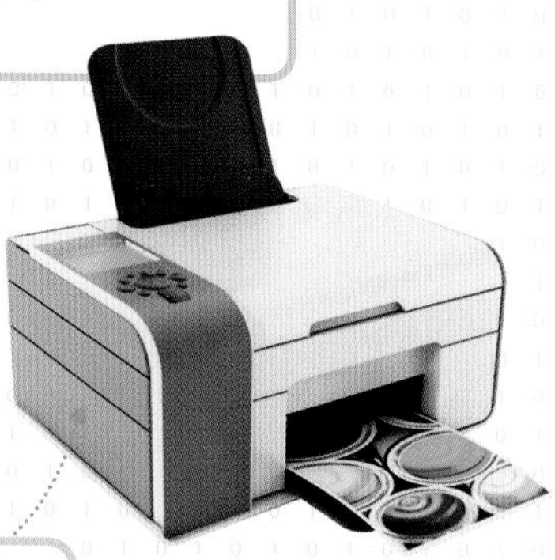

Speakers

Printer

Connectivity is the way in which a computer can communicate with other devices, for example over the internet. Connectivity can be through wires, or by a wireless connection. Wireless connections are made using radio waves or similar signals.

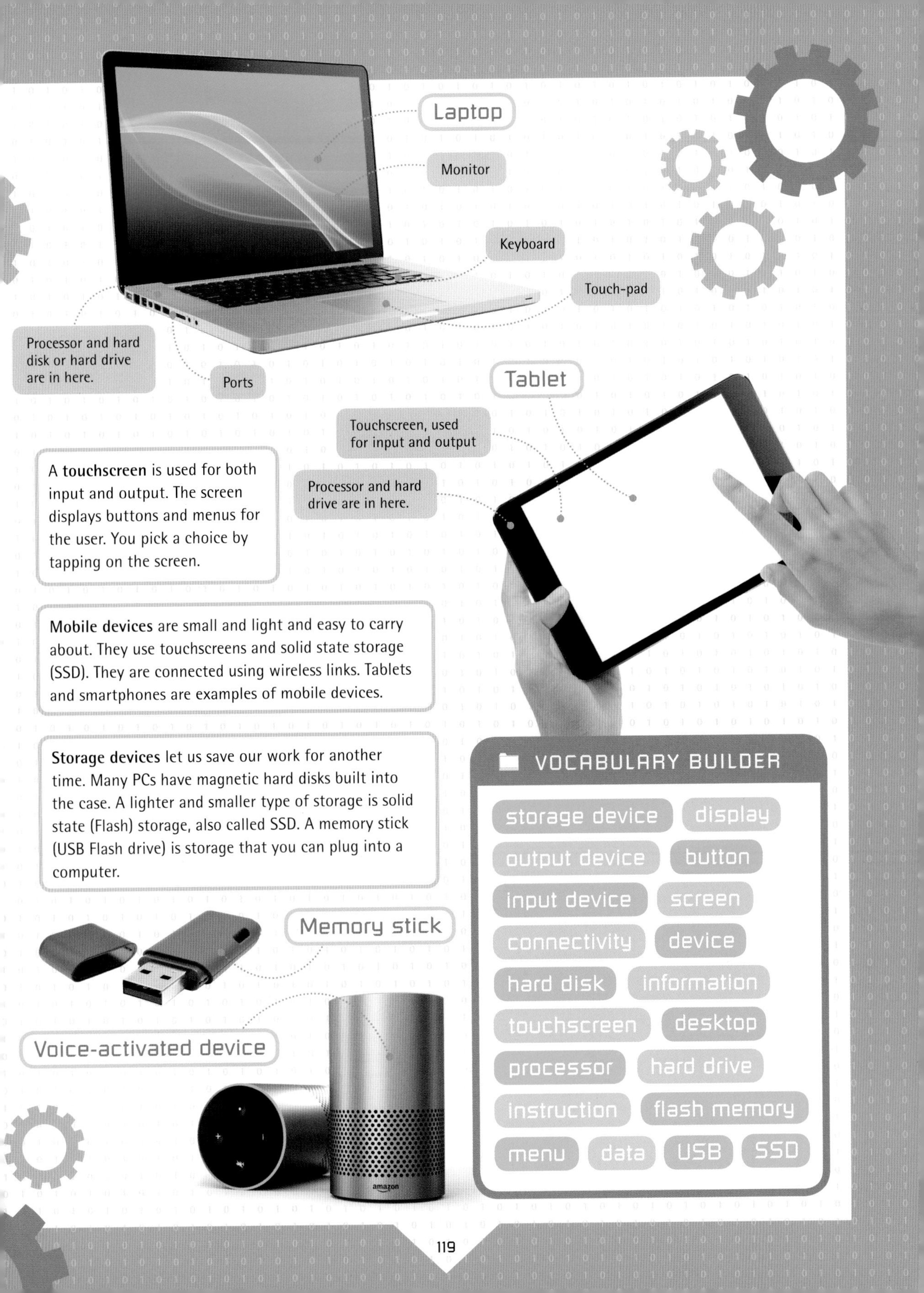

A **touchscreen** is used for both input and output. The screen displays buttons and menus for the user. You pick a choice by tapping on the screen.

Mobile devices are small and light and easy to carry about. They use touchscreens and solid state storage (SSD). They are connected using wireless links. Tablets and smartphones are examples of mobile devices.

Storage devices let us save our work for another time. Many PCs have magnetic hard disks built into the case. A lighter and smaller type of storage is solid state (Flash) storage, also called SSD. A memory stick (USB Flash drive) is storage that you can plug into a computer.

VOCABULARY BUILDER

storage device, display, output device, button, input device, screen, connectivity, device, hard disk, information, touchscreen, desktop, processor, hard drive, instruction, flash memory, menu, data, USB, SSD

The Internet

Every day on Earth more than one billion people use the internet. We use it to play games together, send messages, search for information and shop online. But how does it work?

ONE

The **internet** is a huge network of communication links. All the computers connected to the internet can use these links to share messages and information. One of the main uses of the internet is to share web pages. The collection of all the web pages in the world is called the **World Wide Web**. To show how the internet works, let's suppose that you want to see some red shoes from *bigshoeshop.com*.

TWO

Bigshoeshop is an **e-commerce** company. They use the internet to sell their products. They put information about their products onto a **web server**. A web server is a computer with a permanent connection to the internet. Anyone with an internet connection can **download** a web page from the server to their own computer.

THREE

The software we use to look at web pages is called a **web browser**. At the top of the web browser window is an address bar. The address of a web page is called a **URL** (Uniform Resource Locator). You could type the URL *www.bigshoeshop.com* into the address bar. Or you could click on a **link** from some other internet page such as **Google**. Either method will attempt to make a connection to the web page.

FOUR

Your web browser sends out a message to *www.bigshoeshop.com*. The message is passed along a chain of computers called **routers**. With each transfer the message gets closer to the right web server. The message is a request to see the web content. The message has a header which identifies your return address.

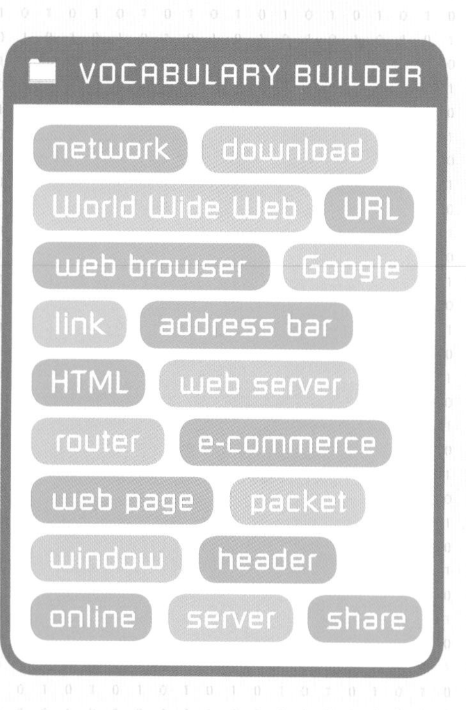

FIVE

When it receives the message, the web server sends the web content back to your address. The content may include text, images, sound and video. The content is split into sections called **packets**. Each packet makes its own way to your computer. Your computer receives all the packets. It puts them back together in the right order.

SIX

The web page is defined using a language called HTML (Hypertext Markup Language). Your web browser can understand HTML. It reads the HTML and uses it to build the web page on your screen. Now you can see the *bigshoeshop* page with all the information about shoes. You might want to buy the shoes. If you do then you send another message back—and so the internet conversation continues...

Programming in Scratch

Scratch is a **programming language** designed to help anyone, but especially young learners, to learn to write their own **programs**. It is free to download from *https://scratch.mit.edu/* and has been used by people in more than 150 countries around the world.

Instead of typing text, you build programs by fitting blocks together. Each block stands for a **command** or **program structure**.

A Scratch program can control the movement of an object on the screen. This object is called a **sprite**. The first sprite you see when you start Scratch is a cat.

A Scratch program begins with an **event**. This program will begin if you click on the sprite. The program produces output. The sprite says 'Hello' and makes a 'meow' sound.

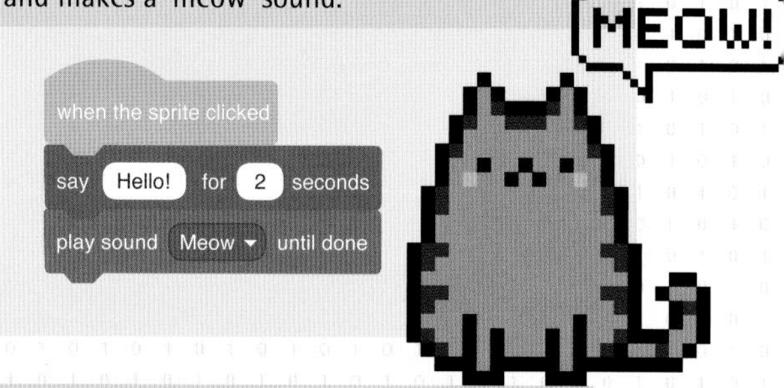

This Scratch program makes the sprite move about the screen. The movement blocks are inside a 'forever' loop. That means the sprite will move about 'forever' (at least until the program stops).

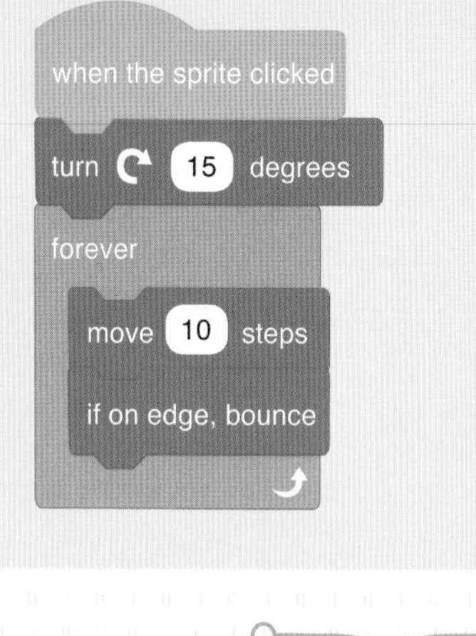

This Scratch program asks the user how many boys and girls are in their class. The user **input** is stored as two **variables** called boys and girls. The two values are added together using an **operator** (the addition sign). The result is output by the sprite.

A spaceship computer game

This is part of a program that makes a simple computer game. The sprite is a spaceship.

The program includes a **logical test**. It tests if the sprite touches a star.

The commands that follow are only carried out if the test is true. The ship jumps to a random location and the player loses one point.

A structure like this is called a **conditional structure.**

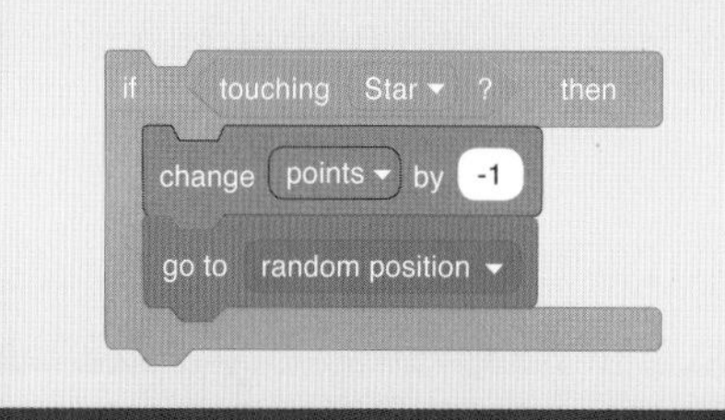

Here is what it looks like when you play the spaceship game.

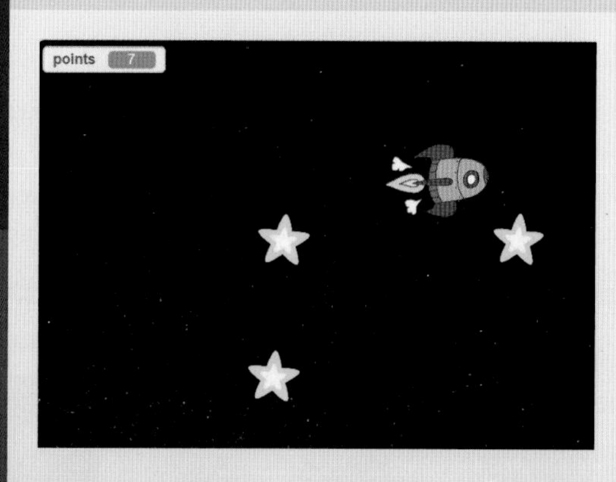

This is the complete spaceship program. What program features can you spot?

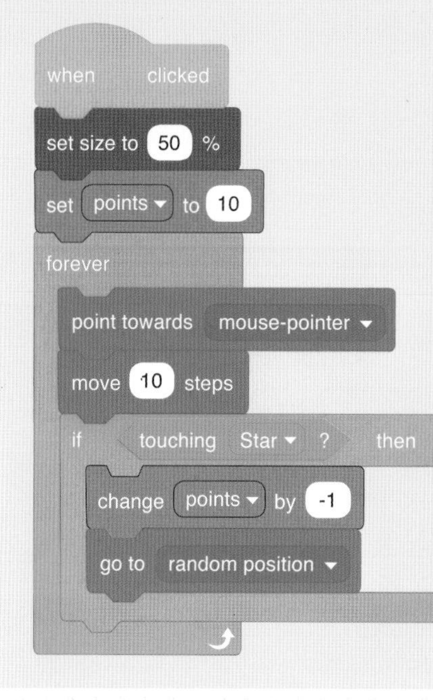

VOCABULARY BUILDER

programming language
conditional structure
program structure
logical test
sprite
if statement
output
command
program
operator
variable
loop
input
event
mouse pointer
block
object
screen
value

Programming in Python

Python is a **text-based** programming language that is designed to make programs clear and easy to read. Python is good for learning programming techniques. The standard version of Python does not have the lively multimedia features you will find with Scratch.

You make programs in Python by typing commands using letters, numbers and other characters and symbols.

Python is not just for learning programming. Many real-life systems, games and applications have been made (or partly made) using Python. Examples include YouTube, Dropbox and Spotify.

Python is managed by the non-profit Python Software Foundation.

Python is available for free download from the Python website.

https://www.python.org/

Click on the 'Downloads' button to **download** Python onto your own computer.

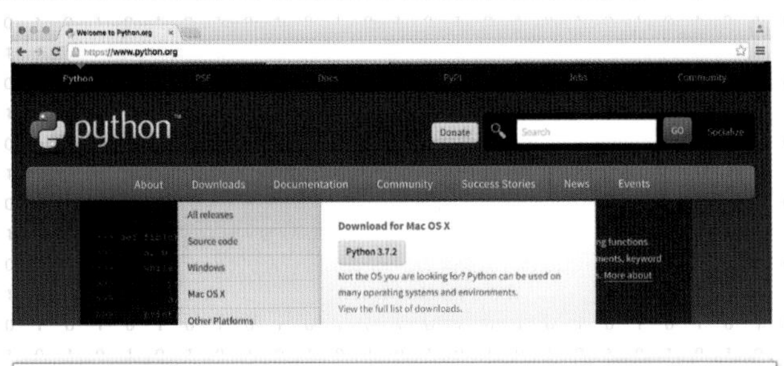

The Python **IDE (Integrated Development Environment)** lets you make and run Python programs. Type a Python **command** and press Enter to see the result right away. Or open a new file and create a **program**.

Python 3.6.0 Shell

File Edit Shell Debug Options Window Help

```
Python 3.6.0
>>> print("Hello World")
Hello World
>>>
```

Ln: 5 Col: 4

A **variable** stores a value. This program shows the use of variables.

In this example

1. The variable ticketprice is **assigned** the value 5.99
2. The user **inputs** a number. The input is assigned to the variable howmany.

Python 3.6.0 Shell

File Edit Shell Debug Options Window Help

```
ticketprice = 5.99
howmany = int(howmany)
```

Ln: 3 Col: 4

Programmers often add **comments** to their programs. In Python a comment is shown by the symbol #.

Comments are ignored by the computer. They are added to help the human reader. They make the program more understand-able.

Python 3.6.0 Shell

File Edit Shell Debug Options Window Help

```
ticketprice = 5.99

#input number of tickets
howmany = input("Enter how many tickets: ")
how many = int(howmany)

#calculate total cost
total = ticketprice * howmany
print("The total price is: ",total)
```

Ln: 7 Col: 4

When the program is run the user enters the input. The program processes the input to produce results and displays these on the screen.

A **conditional structure** is controlled by a **logical test**.

This Python program shows a conditional structure. An exam mark is input to the computer and converted to an integer from a string.

The logical test uses the **relational operator** > (greater than). It tests if the mark is greater than 50.

1. If the test is true the message 'you passed' is shown.
2. If the test is false the message 'take the test again' is shown.

A **loop** structure repeats. This Python program shows a **condition-controlled** loop. In Python condition-controlled loops begin with the word **while** followed by a logical test.

The input command asks the user to enter the password. The command will repeat until the user enters the right password.

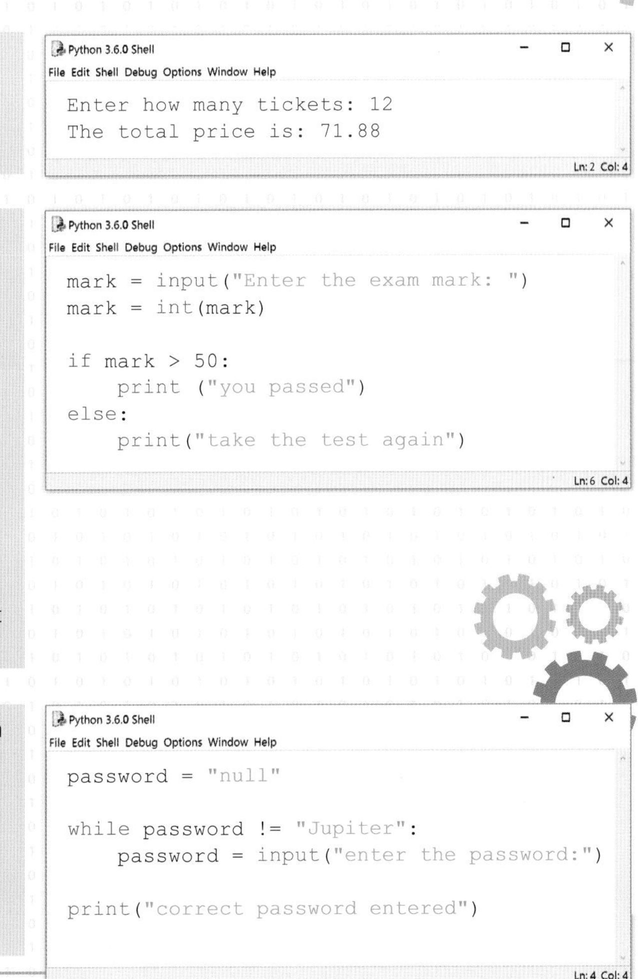


```
Enter how many tickets: 12
The total price is: 71.88
```

```
mark = input("Enter the exam mark: ")
mark = int(mark)

if mark > 50:
    print ("you passed")
else:
    print("take the test again")
```

```
password = "null"

while password != "Jupiter":
    password = input("enter the password:")

print("correct password entered")
```

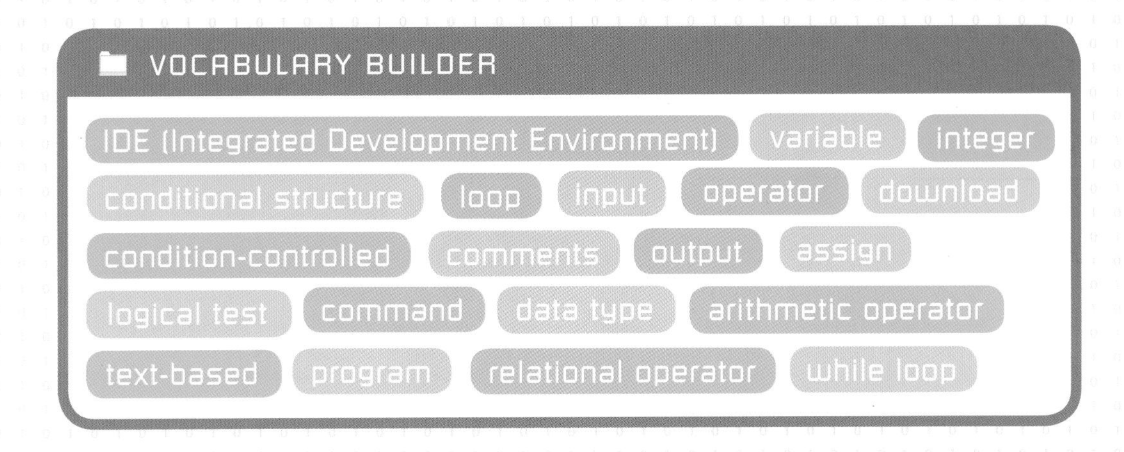

Computer Games

Computers and mobile devices can be used to play computer games. You can also buy specialist **consoles** that are only used to play games. Computer games are a multi-billion dollar industry. A popular game can have millions or even billions of players. Some games are played professionally in worldwide tournaments for millions of dollars in prize money. They are known as **eSports**.

Games are organised into **genres**. Games in the same genre have similar features. This means you have an idea what a game is like before you buy it. It means you can find more games of the type that you like.

ACTION

Action games have fighting and shooting. When you play this type of game you usually move about in a landscape. You are at risk from dangers. You often have to battle enemies or monsters. You may play using the **keyboard** or another type of controller. Some games of this type are unsuitable for children.

ADVENTURE

Adventure games typically involve exploring an **online** world. Adventure games are usually less violent than action games. Games in this genre give the player a varied range of challenges. You may beat the challenges by solving problems, by collecting useful resources or by working with others. Adventure games are often set in a colourful fantasy world.

SIMULATION

Simulation games are intended to be somewhat like real life. The characters in the game have to do real life things like find jobs and get married. Or they might do a realistic activity like flying a plane. A form of simulation is **VR (virtual reality)** which gives you a realistic **3D experience**. This may involve putting on a **VR headset** that shows you a 3D image of the game world.

PUZZLE

A puzzle game is not as realistic as other game genres. The design is colourful and bright. You may have to fit shapes together, find matches or use logic to solve puzzles. Puzzle games are often very quick to play. They are suitable for busy people to play on a **smartphone**.

MOVEMENT

Some games mean the player has to move about in the real world. A good example is some games that are played on the Nintendo Wii. This console has **sensors** that can detect your movements. So to play these games you mime actions like playing tennis or dancing.

MMORPG

MMORPG stands for *massively multiplayer online role-playing game*. Players all over the world can **log on** using an **internet** connection. Information about all the players and their actions is **hosted** by the game company. The players can interact in a shared game.

SANDBOX

A **sandbox** game is one where there are no rules or aims. Instead the game gives you a range of interesting resources. You can use them to do what you want. You might make buildings, explore a landscape or look after animals and crops. These games are good for creative play. They can also be used in the classroom for learning through play. For example, you might build a castle with its different features to help bring history to life.

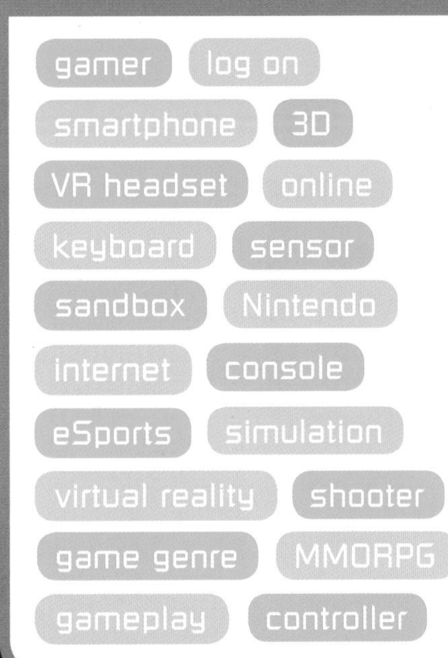

Perfect for use with:

age 9+ age 10+

age 11+ age 11+ age 11+

For more vocabulary and language skills:

age 8+ age 8+ age 8+ age 8+

FOOD MILL A good-quality food mill with different-sized mesh discs can be used to sieve fruit to extract the purée, which is excellent for making jams and gives a pleasing texture, without 'bits'. You can also purée apples without having to peel and core them first.

JAM FUNNEL This is essential for pouring hot jam safely into jars. Choose one small enough to fit into most of your jars but wide enough not to become clogged with pieces of fruit. Sterilize and warm the funnel in the oven along with the jam jars. A sterilized, warmed scoop is useful for ladling jam into the funnel.

JAM THERMOMETER Although not essential, this is useful for testing for setting point. Choose one that goes up to at least 110°C/230°F and has a clip to attach it to the side of the pan.

JELLY BAG Ready-made jelly bags in a plastic stand that will fit over a bowl are ideal for straining the juices from cooked fruit, but you could make your own using muslin, nylon, a clean tea towel or calico tied across the legs of an upturned stool.

MUSLIN You will need squares of muslin to hold pips, stones and spices that require cooking in with jams and chutneys. A generous square of fabric can be gathered together around the ingredients and tied into a bag with natural string or twine. Alternatively, you can buy little drawstring calico bags made specifically for this purpose.

WAXED PAPER DISCS AND CELLOPHANE COVERS
Hot jam should be covered with waxed discs, placed waxed side down, immediately after bottling before the screw-top lids are put on. These help to keep the preserve sterile. The discs can be bought to fit 450g (1lb) and 900g (2lb) jars. Choose discs that will completely cover the jams, jellies and marmalades right to the edges where possible. The discs are often sold in packets that also contain cellophane circles, elastic bands and labels. If you have any jars without lids, these cellophane covers can be used as lids.

LABELS Label all your preserves so you know how long they have been stored. Chutneys and pickles, which benefit from a maturing period, also need to be labelled with this information.

preserving techniques

There are two basic methods for making jam. The traditional method involves cooking the fruit before adding the sugar and boiling to setting point. The macerating method requires leaving the fruit and sugar, which have a lower sugar-to-fruit ratio, together, ideally overnight, to draw out the juices and intensify the flavours, before boiling to a set. This method gives a softer-setting jam with a more syrupy consistency.

JAM: THE TRADITIONAL METHOD

COOKING THE FRUIT Place the fruit in a preserving pan with some water, the quantity of which will vary with the type of fruit. Soft fruits may not need any water, as they will quickly break up and release their juices when heated and mashed with a spoon. Harder fruits, however, will definitely need some water and a longer cooking time to soften them and release the pectin and acid. Simmer the fruits gently. Blackcurrants and plums should be cooked until their skins are soft. Once sugar is added, skins may become tough if they haven't been cooked enough before.

ADDING THE SUGAR Use white granulated sugar, or preserving sugar if the fruit needs more pectin. The amount of sugar needed varies, but the minimum amount recommended for 1kg (2lb 4oz) fruit is 700g (1lb 9oz) sugar. Ideally, use 750–850g (1lb 10oz–1lb 14oz) sugar for a softer-set jam and 1kg (2lb 4oz) sugar for a traditionally prepared jam.

It is best to warm the sugar in a bowl in the oven (about 20 minutes on its lowest setting) before adding it to the fruit, as this will help it to dissolve quicker. Take the jam off the heat and allow it to cool slightly so it isn't still boiling when the sugar is added. Stir continuously over a low heat until the sugar has completely dissolved. (If the jam comes to the boil before the sugar has completely dissolved, it may crystallize during storage.)

BOILING THE JAM When the sugar has dissolved, turn up the heat and bring the jam to the boil. How long it needs to boil for will vary and this is something that becomes more apparent with experience. The jam needs to maintain a high temperature to reduce and thicken so it will set as it cools. Sometimes this can take a matter of minutes and on other occasions up to half an hour, depending on how much water the jam contains. This stage is often referred to as a rolling boil.

TESTING FOR SETTING POINT After 5–10 minutes of rapid boiling, test the jam to see if it has reached setting point. This can be done in several ways (see below). Remove the pan from the heat while testing, so the jam doesn't over-cook.

The cold plate test Put a small plate in the freezer to chill beforehand. Drip a pool of jam onto the plate with a spoon and allow to cool for a few seconds, then draw your finger through the jam. If setting point has been reached, the surface will wrinkle. When you raise your finger from the plate, the jam will form a strand rather than dripping off.

Using a jam thermometer
Dip the thermometer into hot water, then push it into the jam, preferably in the centre of the pan. If the temperature reaches 105°C (220°F), setting point has been reached.

The flake test Dip a wooden spoon into the jam, then hold it above the pan. Leave it to cool for a few seconds, then let the jam fall off the spoon back into the pan. If the jam has a gloopy consistency and forms strands or flakes that hang onto the spoon, setting point has been reached.

If setting point has not been reached, place the pan back on the heat and continue to boil rapidly, testing again at 5-minute intervals.

SKIMMING During boiling, a scum sometimes forms on jam, jelly or marmalade due to bubbles rising to the surface. This scum is harmless but can spoil the appearance of the preserve. Stir in a small knob of butter to help disperse the scum or use a metal spoon to scoop it away.

DISPERSING THE FRUIT Whole fruits or large pieces of fruit often rise up to the top of a jam, and, as the jam sets, they are likely to stay put. If you are making a softer-set jam with pieces in, you may have to live with this, but for a thicker set, leave the jam for 5–10 minutes prior to potting up, then stir to distribute the pieces evenly.

POTTING AND STORING Have your hot jars and jam funnel ready. Pour the jam into the jars, filling them to the top. Cover them straight away with a waxed disc pushed down onto the surface of the jam and then with a screw-top lid. Leave the filled jars to cool. Jars filled to the brim can also be left to cool upside down, as this helps to produce a vacuum as it cools. Store cooled jam in a dry, cool larder or cupboard. Note that some preserves are left to go cold before being sealed – see individual recipes.

JAM: THE MACERATING METHOD

Allowing fruit and sugar to macerate together before cooking draws the moisture and juices from the fruit and preserves the flavour. Cooking times are reduced and flavours intensified, and it is possible to use a lower sugar content (see page 7), depending on the sweetness of the fruit.

Prepare the fruit as directed and place in a ceramic, glass or stainless steel bowl. Add the sugar, cover with a plate or push a piece of greaseproof paper down onto the surface of the fruit to hold in the moisture and leave to macerate: softer fruits for 6–8 hours and harder fruits for up to 36 hours. You will see the sugar soak up the juices and a considerable amount of liquid begin to dissolve the sugar.

Pour the fruit into a preserving pan and stir over a gentle heat until the sugar has completely dissolved. Occasionally, the mixture is left to macerate again, but if not, bring the jam to setting point as for the traditional method and pot up.

MAKING MARMALADE

If making citrus marmalade, it is important to cook the citrus rind properly, and this can take 1½–3 hours depending on the method you choose. Poaching the oranges whole is my preferred method, but removing the rind and shredding it at the outset is an alternative. If you use waxed fruits, you will need to scrub them before use, but just a rinse will do for unwaxed ones.

POACHING Wash the whole fruits and place in a heavy, lidded casserole with a tight-fitting lid. Pour in enough water to just cover the fruits, so they begin to float, then cover and place in a preheated oven, 180°C/350°F/Gas Mark 4, to poach for 2½–3 hours, by which time the skins will be softened. Leave until cool enough to handle, then, using a spoon, lift the fruits from the liquid, halve them and scoop out the insides, gathering together all the pith and pips and collecting any juice. Slice the rind into strips. Place the pips and pith in a square of muslin and tie into a bundle with string. Pour any collected juice back in with the cooking liquid.

PARING THE RIND FIRST Cut the uncooked fruits in half and squeeze out and collect the juice. Save the pips. Pare the rind and chop into thin shreds. Chop the pith finely. Place the pips in a square of muslin and tie into a bundle with string. Place everything in a pan and add enough water to cover, then leave overnight to soak. The next day, bring to the boil, then simmer for about 1½ hours until the rind is softened and cooked through. Remove the bundle of pips.

From now on, the method is the same whichever form of preparation you used. Add warmed sugar and stir to dissolve, then complete in the same way as if making jam. Once setting point is reached, leave the marmalade for 10-15 minutes, then remove the muslin bag (if using the poaching method) and stir to distribute the rind shreds before potting up.

MAKING JELLIES

A jelly is similar to a jam but does not contain pieces of fruit. The fruit is cooked with water, then poured into a jelly bag and allowed to drip through. Only the juice is made into a preserve. Fruits most suited to jelly making generally have a high pectin content (see page 7).

When making jellies, cook the fruit first with water until tender. You can mash the fruit with a spoon at this stage before pouring it into a jelly bag suspended over a container to catch the drips. For the clearest jelly, allow plenty of time for the juice to drip through (overnight is ideal), and don't squeeze the bag, as this will make the jelly cloudy. (It is often possible to re-boil the contents of the jelly bag using half the original amount of water and pour it through the bag again to get the maximum amount of juice and pectin from the fruit.)

Now measure the juice to work out how much sugar will be needed. The general rule is 450g (1lb) sugar to every 600ml (1 pint) juice, which will make roughly 750g (1lb 10oz) jelly. Place the juice in a preserving pan and add warmed sugar as for jam-making, stirring until completely dissolved. Bring to a rapid boil and cook on a high heat to reach settting point, as before.

Rather than waste the residue in the jelly bag, push it through a food mill, collect the purée, and sweeten it, to use for pie fillings or to make into another preserve.

MAKING CURDS

Fruit curds have shorter keeping times than jams and jellies: just 2 months unopened in a cool place. Make them in small jars, as once opened they should be kept in the fridge and eaten within 2 weeks. You can also pour curds into suitable containers and freeze for 6 months.

Curds are made using sugar, butter and eggs and so are more like a custard than a jam. They are best

suited to tart, fruity flavours. So that the eggs don't curdle or cook on too high a heat, it is best to use a double boiler or a basin set over a pan of simmering water. You may have to stir the curd continuously for 20–30 minutes until it thickens and will coat the back of the spoon, but the result will be worth the effort.

Prepare a fruit purée first by cooking the fruit in the minimum amount of water, if any at all, until tender. Softer fruits such as raspberries and blueberries require hardly any cooking, while gooseberries and squash need more. Press the fruit through the fine disc of a food mill or a sieve and collect the purée.

MAKING CHUTNEYS AND PICKLES

Chutneys can be made from fruits and vegetables mixed with vinegar, sugar and spices. They are easy to make, generally just requiring all the ingredients to be thrown together in a preserving pan and cooked for a couple of hours. it is important to pot up in jars with vinegar-proof lids.

Try to leave chutneys and pickles to mature for 6–8 weeks or even a few months before eating, as the flavours mellow over time.